GW00939370

STUDIES IN
QUALITATIVE
METHODOLOGY

Volume 2 • 1990

REFLECTIONS ON FIELD EXPERIENCE

STUDIES IN QUALITATIVE METHODOLOGY

A Research Annual

REFLECTIONS ON FIELD EXPERIENCE

Editor: ROBERT G. BURGESS
Department of Sociology
University of Warwick

VOLUME 2 • 1990

Greenwich, Connecticut *London, England*

55 Old Post Road, No. 2
Greenwich, Connecticut 06836

JAI PRESS LTD.
118 Pentonville Road
London N1 9JN
England

ISBN: 1-55938-023-3

Manufactured in the United States of America.

CONTENTS

PREFACE

The papers by Jacquelin Burgess and Carolyn Harrison, Alan Prout and Helen Rainbird were originally presented at a series of seminars that I organized on Field Research held at the University of Warwick, Coventry, England with the support of the British Economic and Social Research Council to whom I am most grateful. I am also indebted to Pat Langhorn for her splendid secretarial support and hard work in preparing this material for publication.

Those readers who wish to obtain further details of these volumes or who have material for consideration should contact me at the Department of Sociology, University of Warwick, Coventry CV4 7AL, England.

Robert Burgess

INTRODUCTION

The main purpose of *Studies in Qualitative Methodology* is to provide a forum for investigators working on "ethnography" or "field research" or "case study" to take up issues and debates concerned with methodology; the relationship between data collection and data analysis; the relationship between theory and method and the implications of qualitative research for policy and practice. All of these issues are covered by the papers in the present volume where researchers have been invited to reflect on the principles, processes and problems associated with a particular project or a range of qualitative projects on which they have worked. The result is a series of first person accounts that discuss different dimensions of field experience.

This volume, therefore, follows in a substantial tradition of autobiographical writing in Britain, the United States, and Australia covering a range of disciplines. Among some of the better known examples are *Sociologists at Work* (Hammond 1964); *Doing Sociological Research* (Bell and Newby 1977); *Fieldwork Experience* (Shaffir, Stebbins, and Turowetz 1980); *Doing Feminist Research* (Roberts 1981); *Social Researching* (Bell and Roberts 1984); *The Research Process in Educational Settings: Ten Case Studies* (Burgess 1984); *Doing Sociology of Education* (Walford 1987); and *Doing Research in Organisations* (Bryman 1988).

Each of these volumes involved researchers in producing a first person account of their research experiences. Many provide very descriptive accounts. Some provide little detail about the relationship between theory (in their appropriate discipline) and method and the way in which it related to their research experience. Accordingly, a central feature of this volume is a discussion of the relationship (as appropriate) between theory and field experience, method and field experience, and the relationship (where appropriate) to policy and practice. In this way, it is intended that theoretical and technical issues should become an integral part of the account that critically examines some of the principles associated with the research process.

When I edited *The Research Process in Educational Settings* (Burgess 1984) I gave authors a list of questions that they could consider in relation to their project. These questions proved useful and were, therefore, given to the authors in this volume to help them think about the theoretical and technical issues that could be included in their accounts. The questions were:

1. What was the origin of the project?
2. What were the research aims and objectives? What key questions were addressed? How were they developed? When were they developed? Why were they developed?
3. What form did research design take? What are the technical procedures? How were the procedures modifed by research practice?
4. How was access obtained? Who were the sponsors? What were your relationships with sponsors and gatekeepers? How did these relationships influence data collection and analysis?
5. What sampling procedures were adopted? How were they used? Why were they used? How were sampling procedures modified?
6. What major groups and events were studied? Who were the main informants? How were they selected? What was their role in the project?
7. What methods of social investigation were used? Methods of observation, participant observation, conversations, informal interviews, documentary methods, unobtrusive measures. What relationships existed between the methods? How were the methods modified in relation to the project?
8. What form did field relations take? What roles were taken in the project? What were your relationships with the informants? What was the influence of sex and gender on field relations? In the case of research teams, what were the relationships among team members? What were the power relations on the project?
9. What language skills were required in data collection and data analysis? What linguistic skills were developed in the field and how important was this for field relations?
10. What data recording procedures were used? What methods were used for recording, organizing and filing field notes and keeping field diaries?
11. What was the relationship between data collection and data analysis? What were the informal processes involved in data analysis? What technical procedures were used for analyzing field data? What form did the writing up take?

12. What was the role of theory in the project? What was the relationship between theory, data collection, and data analysis? What form did theorizing take?
13. What ethical problems were confronted in the project? How were these problems handled?
14. What form did data dissemination take? What was the impact of the project on sociology, educational studies, policy and practice and so on?
15. In what ways would you develop further projects in this area? What methodological advice would you offer?

However, it was stressed that this was not a checklist. Indeed, authors were not required to write about all these issues. Instead, they were invited to use some of these issues in reflecting on their field experience and discussing the relationship between theory and research, and methods and research, as well as the way in which their study was conceived, funded, managed, conducted, written, and published. While this was intended to help maintain some continuity across the contributions to this volume, it was not intended to bring a uniformity of style. Instead, contributors were encouraged to write in a way that would highlight aspects of their own field experience.

OVERVIEW

This volume contains a range of papers that focus on research experience. The contributors come from a range of disciplines: anthropology, sociology, social policy, geography, health studies, management and business studies, and education. Yet all their papers demonstrate a range of similar issues concerned with the conduct of qualitative research.

The first paper by Stephen Fox is based on the first major piece of qualitative research conducted by this researcher for a Ph.D. Fox was studying part-time MBA students and discusses some of the key phases of his research experience: gaining access, principles of selection and issues concerned with data collection and analysis. A key feature of his paper is concerned with learning to use ethnomethodology in the course of collecting and analyzing data in this study.

Some of these themes are taken up in subsequent papers. For example, Janet Finch and Jennifer Mason are concerned with issues of selection and sampling in their study of family obligations. In particular, they are concerned with the way in which the principle of theoretical sampling was applied and developed. An important dimension of their paper consists of extracts from a collective diary which offers insights into collaborative work as well as highlighting different dimensions of decision making in qualitative research. Among the questions they address is: how do you select informants?

It is this issue that is also taken up by James Henslin in his studies of homelessness. Here, Henslin demonstrates how a sociologist conducts work in an unfamiliar setting in the society in which he is a member. However, Henslin is not just concerned with who is researched but also how they are researched and where. Similar themes are also taken up by Helen Rainbird, a social anthropologist, who demonstrates how such themes arise in research conducted in a society other than your own. In addition, she also draws attention to the ways in which gender influences where research occurs, how it is conducted and with whom.

A similar range of themes are also picked up by Kristine Mason who turns to a study of rural England using an anthropological approach. She illustrates the way in which her role as a participant observer helped her to become located in the locality through the playgroup, the youth club and the local council. It is through her observations that she was able to generate theories concerned with girls' schooling in a rural context. However, central to this gendered analysis of schooling is the gender of the researcher as Mason demonstrates how being a woman and a mother helped to facilitate her fieldwork.

Similarly, Joan Chandler's research on women who were married to Royal Navy personnel in the United Kingdom also takes up questions of gender and fieldwork. She illustrates the way in which her research experience assisted her in making an analysis of situations where a woman researcher works with women. She concludes that researchers need to be aware of their place within fieldwork and in the process of data analysis.

The evaluation of different kinds of research methodology is a central theme in the paper by Jacquelin Burgess, Barrie Goldsmith, and Carolyn Harrison who comment on the tensions that existed among members of the team who had interests in a range of methodologies. In addition, their paper raises questions about the difficulties associated with team-based reasearch and the relationship between qualitative research, policy and practice. It is the relationship between reserarch and practice that is also followed up by Alan Prout in a paper that draws on two projects to explore the practical implications of research on parenthood education and on sickness absence in a primary school class.

In the two papers that follow, the researchers demonstrate the lessons that can be learned by comparing research experiences on several field projects. Norman Friedman reflects on three ethnographic studies where he was a high school substitute teacher, a Hollywood actor, and a religious school supervisor. Here, he takes up Prout's theme of the links between research and practice as well as comparing different styles and strategies of investigation in ethnographic research. In a similar way, Virginia Olesen uses reflections on her field experience in three projects to examine the link between fieldwork and policy and fieldwork and theorizing. Her analysis of three studies will also help readers to make links with some of the other themes raised in this volume concerning research sponsorship, research funding, and the role of research teams in qualitative studies.

The papers that are included in this volume highlight the "pay off" of reflectivity and periods of reflection in the conduct of qualitative research.

Robert Burgess
Series Editor

REFERENCES

Bell, C. and Newby, H. (eds.)
1977 *Doing Sociological Research.* London: Allen and Unwin.
Bell, C. and Roberts, H. (eds.)
1984 *Social Researching.* London: Routledge and Kegan Paul.
Bryman, A. (ed.)
1988 *Doing Research in Organizations.* London: Routledge and Kegan Paul.
Burgess, R.G. (ed.)
1984 *The Research Process in Educational Settings: Ten Case Studies.* Lewes: Falmer Press.
Hammon, P. (ed.)
1964 *Sociologists at Work.* New York: Basic Books.
Roberts, H. (ed.)
1981 *Doing Feminist Research.* London: Routledge and Kegan Paul.
Shaffir, W.B., Stebbins, R.A., and Turowetz, A. (eds.)
1980 *Fieldwork Experience: Qualitative Approaches to Social Research.* New York: St. Martin's Press.
Walford, G. (ed.)
1987 *Doing Sociology of Education.* Lewes: Falmer Press.

BECOMING AN ETHNOMETHODOLOGY USER: LEARNING A PERSPECTIVE IN THE FIELD

Stephen Fox

In this chapter I would like to principally describe how, in the course of a longitudinal ethnographic study of part-time executive MBA students at Manchester Business School, I learned to use the perspective of ethnomethodology to produce analytic effects. The purpose of this research was to illuminate the experience of part-time masters students of business in a conventional business school setting.

In the process of starting my research and negotiating access, I faced a variety of hurdles; micro-political as well as methodological that this chapter documents. One of these was the prevailing norm within the business school for research to be functionally oriented rather than interpretive and exploratory. The first three short sections of the chapter present an autobiographical account of becoming registered for a Ph.D. and developing access. The fourth section seeks to illuminate the way in which a theoretical

Studies in Qualitative Methodology, Volume 2, pages 1-23.

ISBN: 1-55938-023-3

perspective (ethnomethodology) infused my field experience including data collection and analysis, through extended reference to Becker's (1963) study: "Becoming a Marijuana User."

AUTOBIOGRAPHICAL CONCERNS BEHIND THE RESEARCH

My research proposal was completed in May 1984, eight months after I began field researching my focal culture of part-time masters students of management. While this was personally significant in terms of getting myself registered as a doctoral candidate, it is also indicative of some of the problems ethnographers face when doing "exploratory" field studies. The proposal included not only a general statement of the research problem and why it was "relevant" to the practical concerns of management educationalists and learners, but a literature review and an early outline of my methodology, *but also* "Emerging Themes and Methodological Issues." I felt constrained to show that after only eight months "in the field" already my "high risk" strategy was "paying off"—field reserach was seen as "high risk" because insights could not be programmed into the methodology whereas chi squareds, confidence limits, and probabilities could be in "lower-risk" positivistic research designs.

Presenting my proposal in this way was largely because of the way I saw the institutional constraints of the business school at which I was registered, and also, I surmised, due to a culture in which "pay-offs" or "likely pay-offs" had to be in the justification for anything.

However, long before writing the May 1984 proposal I had already chosen my "research problem" in a broad way and my preferred methods. My interest in management education stemmed from my own experience as an undergraduate in business studies. I had taken an H.N.D. and a B.A. in business studies at a polytechnic where practical emphasis was supposedly emphasized. However, I had been somewhat skeptical about the practical value of much that was taught, on two main counts. In the first place, my father had started a small business, in partnership, when I was eight years old. We moved to raise the capital via a second mortgage. When I was sixteen he folded up the business and worked as a sales representative for a year before beginning a second, this time solo, enterprise operating from home, which he gradually built up. From the start he brought his work home with him and my family life was structured around the business and its substantive concerns permeated the discussion between my parents. I was taken to his premises as a boy and learned several productive processes as well as the need to keep books and manage a small unionized labor force, other directors and customers. In many ways running a small business was part of my everyday reality before entering higher education. The difference between what I learned while doing the

H.N.D. and B.A. and what seemed relevant to my father's everyday concerns seemed striking.

I was not worried personally by the difference; I happily became interested in organizational sociology and analysis and intrigued by epistemological problems, and explained the difference as a result of scale. What is taught in most business studies programs is more relevant to big organizations rather than small.

In the second place, this explanation was eroded when I went as a sandwich student to the penultimate year of my B.A. to work for a profit-center of a large nationalized corporation. I expected the managers here to be much more familiar with the terms used on my courses. But they were not. I worked in the management development and training department of the personnel function and although my main duty was coordinating a retrenchment and voluntary redundancy program for middle-managers, I rubbed shoulders with all the training managers and even helped them produce in-house course designs. Here they did refer to "motivation theory," "McGregor," "Herzberg," and "standard costing," but in doing so demonstrated a surprisingly low appreciation of these theories and techniques.

I returned to my final year as an undergraduate aware of the broad gap between the theories and research findings with which I was familiar and the knowledge and understanding of even senior managers in big business. In my final year, however, I found that my experience of work in industry certainly did enhance my own understanding particularly of organizational analysis and industrial relations.

As I entered business school as a Ph.D. student, I was aware of the broad gap between what the academics knew and what most practicing managers I had ever come across—from small to large business—knew. My own intellectual curiosity was firmly applied to the epistemological and methodological dilemmas of researchers particularly in organizational analysis and sociology.

Initially, I was fascinated by the topic of decision making. After a large literature review of decision making, I was more than ever *undecided* about how to go about researching the topic. However, I had a preference for using case-study methods with participant observation. My preference for the latter had again its origins in my undergraduate studies and experience of work. I had written a thesis on the basis of my work in personnel management that was a case-study of a retrenchment program. I used ethnographic methods of participant observation as well as analyzing many documents, ranging from personal letters and minutes of meetings to works newspapers and national industrial relations agreements. This was with the knowledge of my boss, the manager of training and management development, and the management association representatives with whom I came into weekly contact, although not everyone in the organization knew I was a participant observer.

My B.A. thesis utilized a systems perspective criticized and modified from a reading of Silverman's (1970) action frame of reference, which was applied after the event of actual ethnographic data collection. While the practical focus of my research was clearly middle management retrenchment throughout a year of data collection, the theoretical discipline-relevant problems were only afterwards imported. Doing it this way later allowed me to see the logic in Everett Hughes' apparently scandalous remark:

> that one should select the research problem for which the setting one has chosen is the ideal site! (Hammersley and Atkinson 1983, p. 40).

Partly because my thesis was rated highly, I retained a bias in favor of participant observation, and ethnography, due perhaps to this early affirming experience. At the same time I realized that the methods of ethnography—especially participant observation—held more direct credibility with practicing managers than more high-flown quantitative and elaborately experimental research designs. In their eyes participant observation was closer to simple bald personal *experience:* a "qualification" that is given more credence in the business world than mere academic qualifications, as Thomas (1981) indicates, reinforcing this personal view of mine.

In my literature review on decision making I also explored motivation and learning theory. I felt that each of these areas were concerned with the same thing: how people's courses of action arise producing an on-going stream of decisions, action, and experience running through successively changing settings. I became interested in how people learn in everyday life, and how that learning influences future decisions, action and experience as their lives unfold. I switched my substantive focus to management learning in a formal management educational setting, to explore further the gap between the models and constructs of management academics and the learned understanding of practicing managers.

STARTING MY RESEARCH AND GAINING ACCESS

At the time I began my research I was being supervised by a lecturer at the business school who was interested in "blocks to learning," but who did not have a background in sociology. However, it was through his help and political "sponsorship" within the school that I got access to study the part-time masters program. It subsequently turned out that our perspectives and preferred methodologies were too different for us to work together and I switched to a research fellow with a sociology background for supervision. Her Ph.D. had been an ethnomethodological study of a therapeutic community utilizing ethnographic methods, particularly participant observation. Our methodological preferences were congruent and it was through her that I gained access to the more recent literature and ideas in ethnography as well as an initial interest in the ethnomethodological perspective.

However, because I had gained initial access via my original supervisor, that is, with the program director and the students, I was left with a number of "role-dilemmas" as a participant observer. My first supervisor had ran the opening workshop on the part-time masters program I studied and introduced me as a researcher working with him on "blocks to learning." To give me credibility in my own right, he gave me a slot in the opening workshop to feed some findings back to the part-timers from a psychometric inventory they had completed and which I had analyzed. Also I acted as a tutor/facilitator in several group exercises. While this was useful in enabling me to have a clearly defined role in the eyes of the students, it also gave me some problems when I tried to become more of a politically neutral participant observer. By my association with this lecturer I felt I had become associated with "the faculty" in the eyes of the course members. Therefore, I was someone they might be wary of.

My next meetings with the members was just before Christmas 1983. I arranged to meet half the group for a "free lunch" provided by the school to discuss "how they were finding the course so far." I did this with a psychologist who was a private management consultant employed by the school as a stand-in for my first supervisor who had to be away on that occasion. We repeated this exercise with the other half of the group a few weeks later. Again this was useful in giving me clear reasons for meeting the group but awkward because it made me look as if I was siding with faculty. The understanding was that they could freely criticize the course to me and the psychologist and we would report the criticisms to interested members of faculty while respecting individuals' confidentiality.

It was shortly after these lunches that I switched supervisors and began simply being a participant observer with no official "facilitator" or "evaluator" strings attached. Following this role transition from being a conduit of student opinion to faculty, to being a "participant-as-observer" (Gold 1958; Junker 1960, p. 36) there remained a tendency for part-timers to come up to me during coffee breaks, for example, and give me detailed criticisms of various features of the course such as the late arrival of coffee, the teaching style of lecturers, the relevance and difficulty of the subject matter. While such information was useful, I had to "renegotiate" my role with the part-time MBAs so that they would no longer expect me to pass these views on to faculty in time to change anything. Instead, I attempted to make it understood that I was primarily interested in their experience of the course per se and that it would not be until my Ph.D. was accepted that I would be reporting my "findings" to faculty.

It took me some months before I felt entirely happy with their expectations of my role. While access to do my research had been gained early on with the students and the course director and administrators, I had to negotiate access with each member of faculty who was due to teach on the course as the occasion arose. The program was planned in general but a more specific termly program was used to state what elements of the course would be tackled by whom and when. Because this termly program was subject to changes—for example, when a lecturer could not attend on a particular day for whatever reason—a weekly timetable was also used. This was sent around part-timers a week in advance informing them and me of last minute changes.

The upshot of this arrangement was that I negoitated access on a week-by-week basis with whichever member of faculty was teaching. In general, I took it that if a lecturer allowed me to sit in on one of his/her sessions, he/she would allow me to sit in on the next.

I preferred to do it in this piecemeal way, rather than send a memo around to all faculty advising them of my intentions or indeed arranging early on a spate of interviews with all faculty members likely to teach on the program to arrange access. I feared that these last two options might raise questions at the faculty board of the "*Who* is this Steve Fox character and *what* exactly is he doing?" variety. And since my initial access had been along the lines of me being a conduit of part-timers' criticisms of the course, which has an evaluatory ring to it, I was afraid that some members of faculty might back off, even if I could convince them of my purely ethnographic interest. Hence my approach was to arrange access with those lecturers I got on with anyway and who were teaching early on. Then as weeks went by and the weekly timetable included names of faculty I did not know, I would go and meet them and say that I had already been researching their group of part-timers with the course director's approval and, for example, Professor so-and-so let me sit in on his sessions and seems interested in the project, so would they mind if . . .

This approach worked well, but even then a few lecturers expressed some doubt and one professor refused point blank.

SELECTION STRATEGIES IN MY FIELD RESEARCH

From the time I switched to studying "management learning" rather than "decision making" I wanted to do a longitudinal single case study using participant observation. As the introduction to this chapter notes, one of my prime motivations was to investigate the gap between theory and practice. In electing to study masters of business students I faced several choices regarding location. Being registered in a business school the most convenient option would be to study one of the internal programs. This is what I eventually did, however, the decision to do so was not automatic.

Because I was initially interested in the relation between the academic approach to and practice of management, I decided to study part-time students because they had their feet in both camps. I then sent off for the brochures of all part-time and full-time M.B.A. type programs in the U.K. and studied the structure of such programs. There was no "industry-standard": some involved a full day a week over two years, some over longer; some involved several evenings or one or two afternoons. Some were "action learning" programs involving modules at the educational institutions and long periods at work doing projects. There could be no "representative program" of all part-time management education, I realized. In addition to masters' programs there were D.M.S. and undergraduate programs. Some postgraduate programs allowed the student to, after completing a common first year, opt to tackle either a D.M.S., M.Sc., or M.B.A. depending on their proclivities, abilities, commitment, and resources.

Consequently I made contact with a range of course directors running different types of programs and in addition to gaining access to conduct field research via participant-observation in my own institution, as described in the previous section, I gained access to conduct field research using interviews in these other institutions. I carried out about sixteen interviews with a range of part-time undergraduate, D.M.S., M.Sc., and M.B.A. students of business, management or a related discipline.

I realized that a study *could* be done comparing emerging common themes in the experience of part-time students of management/business and indeed identified some candidate themes to be further explored. However, I was also aware that some work using a slightly different methodology had already been done (Whitley, Thomas, and Marceau 1981). My preference was for participant-observation to examine in greater depth the experience of a single culture of part-time M.B.A. students. Therefore, I decided to drop my research using interview methods with the other institutions and their alumni, and to concentrate on the part-time program within my own institution.

Having decided the broad location, that is, the particular business school, the particular part-time program, and the particular cohort of students (there were at least two cohorts in the school at any time plus third year part-time masters' students who continued to drop in to solicit help with their dissertations), I then had to decide specifics.

The part-time group was about twenty students, and frequently—especially in the first year—they attended class together. I had to decide how many classes to attend, given they were always on every Friday of the week in term time, and were tied to the usual three terms of the academic calendar.

For the first two terms I was managing my "role-change" with my focal group of part-timers and conducting interviews with students from other institutions. I was at the end of the first two terms before deciding to concentrate on one part-time program. Consequently, from there on the number of classes I attended with this group of around twenty increased, although even then I did not attend all of the classes. Within this single field research site were "subsites" (Burgess 1984, p. 6) that I had to choose between. However, most of the time the whole group of twenty was together in class at least in the third and fourth terms. Therefore, the issue of which subsite to choose only arose (a) when part-timers were at lunch, or during morning coffee and afternoon tea or in the bar in the evening; and (b) when part-timers were split up into project-teams to tackle an assignment, in which case they usually retired in their teams to different rooms of the business school.

My general tactic overall was to mix with everyone as equally as possible to avoid being drawn into a clique or creating one. Several factors assisted this tactic. First, because the group was so small it was not difficult over the two years for everyone, including me, to get to know everyone else reasonably well. Second, because the group was together most of the time, in the same room. Third, because each time the group was split into assignment teams by faculty, the subgroup members were mixed up. This was in order to force students to work with a wider range of people with differing personalities, abilities, status levels, motivation, resources to draw upon and so forth. Therefore, it was relatively easy for me to ensure that I joined different subgroups at different times for different purposes without becoming over-associated with any one.

In the last two terms this became more difficult to manage since part-timers were allowed to choose between a series of options and so split into groups which were then relatively long-lasting (e.g., to study market research, creative problem-solving, public sector management, small business). In this event, I joined the largest concentration of part-timers on their option for a few weeks then another group of part-timers on a different option for a few weeks. However, since some of the groups were small, I could not do the same in all cases, but tried to talk more to the people I missed during lunch, coffee, tea, or in the bar.

Over the course of two years many events from the usual to the unusual occurred. Different people became my "informers" at different times appropos different events. Some of the substantive chapters focus on routine matters and others focus on singular occasions. For example, the chapter "Passing Work" focuses on the routine matter: how do part-timers pass themselves off in lectures/seminars in the presence of faculty. This routine matter is examined in the light of particular "routine" events, but these are not "singular occasions" in the same sense as, for instance, a "champagne breakfast" held on the last day of term before Christmas, or the occasion in which a professor delivered harsh feedback to the whole group regarding an assignment, the influence of which was powerfully pervasive throughout the subsequent months. In terms of specific singular occasions, which everybody could see was an unusual and significant event, then certain key actors in the event or by-standers would usually take it upon themselves to explain how they saw it to me. Sometimes they would volunteer to act as my "informant," sometimes I would solicit them to act in this capacity.

Inevitably, the bulk of "what happened" was routine to the members and unnoteworthy in their terms. On such matters I relied—in taking field-notes—chiefly upon my own observations. Occasionally, I drew attention to such matters with one or two part-timers with the result that they became suspicious as to my motives and interests. On other occasions—say in a lecture—I would be the only one taking copious notes. The members would look at me and ask at coffee-break what on earth I found to take notes about during *that* lecture? I would then explain that it was not so much the content of the lecturer's talk, but the routine processes whereby part-timers managed themselves during the lecture. Through such interchanges the part-timers came to understand something of my perspective, and would even discuss my activities in terms of the classic anthropology that "everyone knows" is done among the natives of far away places by intrepid explorers and field researchers.

It is noteworthy, that the situation of classroom ethnography almost ideally lends itself to the task of taking field-notes. I, as participant observer, sat among students listening to the lecture or discussion and, like them, took notes. To the lecturers, I imagine, I represented a model student: attentive, scribbling ferociously, quiet and polite, whereas, to the students I sometimes represented a "swot" since I was taking notes when they felt they should have been but were not. It was irrelevant that my notes were not particularly about the subject matter the lecturer was at pains to communicate. For part-timers the point at issue was that I sometimes appeared to be doing a better job of being a good student than they were. I, therefore, ran the risk of shaming them either (a) into action (note-taking themselves) or (b) into taking reprisal (mocking me for my industry, "swottiness", sycophancy, and so forth). I am sure my ferocious notetaking assisted my access with the lecturers, however.

THE INTERMINGLING OF DATA COLLECTION AND ANALYSIS USING THE PERSPECTIVE OF ETHNOMETHODOLOGY

I began my research of the part-timers with another supervisor. As time went by, we argued; I preferred an interpretive approach, basically he preferred functionalism. No matter how hard I tried I could not persuade him that participant observation was fine on its own. He wanted me to use psychometric inventories and such. We submitted a proposal to a panel of three other members of faculty, one of whom became my eventual supervisor. As it happened none of them accepted my proposal. I felt it was because I had bent over backwards to include a whole hotch-potch of psychometric methods in order to preserve the working relationship I had with my original supervisor. As a result, the proposal looked not so much like a well triangulated research design as a medley of miscellaneous techniques, methods and an incoherent assortment of mutually conflicting justifications.

Both my supervisor and myself were disappointed at our failure to get the proposal through the panel. I went off to discuss the matter in great depth with each of the other panel members. I explained that I wished to do a straightforward piece of participant observation of only one case-study. One of them (who became my second supervisor) said that was perfectly fine, the others hedged and asked about the generalizability of my eventual findings, validity, reliability?

I was confused. I knew it was legitimate to do a participant observation study of one case. Why could they not see it? They said it was "high-risk," they claimed it was an "insight" methodology and insights could not be guaranteed, hence, they were doubtful about recommending it to the faculty board since what guarantee did they have that I could come up with enough insight to get a Ph.D.?

Eventually, the chairman of the committee suggested that I do something to prove I could do it. Before I could be registered as a Ph.D. student, I must go and conduct some depth interviews with masters' students outside this particular business school as well as carrying out my participant observation with the part-time masters' program. In addition, I ought to go and see a leading researcher in management education (Prof. John Burgoyne of Lancaster University) and discuss my proposed methodology with him.

I went to see John Burgoyne in Lancaster. Some time back one of the other panel members, Barbara Rawlings, had suggested I keep a personal diary of the research process, as well as take field-notes based on my participant observation. I liked this idea. We both shared a common interest in stream-of-consciousness writing, so I kept a diary partly in this mode. I recorded in my diary:

> Last Friday I went to Lancaster to talk to John Burgoyne over lunch—Room A.11, Gillow House. Had a pleasant time. We discussed my research ideas—phenomenology rather than positivism—we discussed problems of where do formal learning and natural learning split—if at all . . .
>
> Anyway upshot of these discussions is that he approves of the idea . . . (Diary Notes: Wed. 14/March/1984).

I also arranged some interviews outside the school, taped the conversations and transcribed them. Then I went to discuss them with Barbara Rawlings, the panel member who approved of my intention to do a straight participant observation of one case-study. She had read one of the transcripts, we discussed it at length and with some argument about the epistemology. The meeting was not entirely a happy one, as the following extract from my personal research diary indicates:

> She thinks I talk like a positivist! Gulp! I'm not. 'You *are* Steve.' No I was speaking the part of the person I interviewed—everyday practical reasoning *is* positivist, therefore it

> sounded positivist. I was trying to read myself into the character. Slippery Fox. Barbara wasn't buying. It was the truth. How did she know!? 'When I hear someone talking like a positivist they're thinking positivistically'—she urged (temper, temper) [she] Makes the assumption cues are not being manipulated by me purposefully to sound like the interviewee. Makes the assumption she's infallible. This is hyperbole. 'Nonetheless that's the *'elocutionary force'* [new term] I heard', she said. God knows what she heard—I'm not God, not infallible.
>
> Yes I *am* a positivist, sorry, I try not to be of course—it's innate like cardinal sin—we're born with it—everyday *'natural'* (?) thinking/everyday practical thinking—sinful, totally depraved (1st point of Calvinism: TULIP). Natural: meaning Carnal.
>
> I must try harder, I must try harder. It's fun to work . . .
>
> O.K. all over, left at 12.00 noon went back [to my flat] for lunch. At 2.00 came back for more. Got a paper to read "'K is mentally ill'. The Anatomy of a Factual Account" (by Dorothy E. Smith)—overtones of Kafka. Came away at 3.20. 'Phoned girlfriend. Happy!
>
> At 4.10 went to see my original supervisor. Not there. Eventually got together round about 4.45. Long talk extending into bar up to 8.00 p.m. He's very tired, me likewise.
>
> Oh forgot to say, had asked Barbara if she's interested in supervising this Ph.D.—she said yes although reservations after this morning's strenuous confusions, and reservations also about finishing in two years. A Ph.D. is the most significant piece of work one does in life according to her. (Diary notes: Wed. 14/March/1984).

My encounter with Barbara Rawlings made me question my perspective. I thought my original supervisor was a positivist and I was an interpretivist. According to Barbara I had the right sympathies, but I still talked like a positivist at times. This was unnerving.

I wanted to complete my Ph.D. quickly and join my girlfriend who had at that time left the country and was living overseas. But the point is that much of the intensity of my anxiety over what kind of Ph.D. research to do was fuelled by this personal dilemma. I had several options: (a) forget the Ph.D. and go abroad, (b) stay and do a quick Ph.D. which inclined me toward positivist methods if they would be quicker, or (c) stay and do a Ph.D. I really wanted to do which might take longer (i.e., a longitudinal participant observation). I chose the latter and decided that if Barbara was still keen to supervise it, I would switch to her. She was and I did. We discussed my interview data, she suggested some books (Hammersley and Atkinson 1983; Garfinkel 1967; Becker et al. 1961) and some articles (Smith 1978; Bittner 1974; Zimmerman and Pollner 1971; Ryave and Schenkein 1974; Sacks 1974) and I read some of them immediately, others later.

I rewrote my proposal and, as stated at the beginning of the chapter, I had it accepted in May 1984. I was registered as a Ph.D. student sometime after that.

However, it was from these rather intense beginnings that I became interested in using ethnomethodology as a perspective for analyzing my data; although there was no pressure from my supervisor to do so. Alongside taking field-notes and interviewing I began reading and inevitably my reading influenced my data-collection. I began to see "documentary method" (Garfinkel 1967, Ch. 3) everywhere.

It was not plain sailing. I became captivated and worried by the idea of becoming an ethnomethodologist. I was torn between (a) intellectual curiosity: that is, a strong desire to understand ethnomethodology and to experience it as a way of looking; (b) careerism: with a business studies background and the need to gain employment when my research grant ran out, I had strong tactical doubts about the practical value of using an ethnomethodological perspective; (c) hedonism: ethnomethodology seemed interesting and illuminating—a revelation if only I could get to see through its lens.

I chose (a) and (c), which were for me closely related, but I had doubts. Once again my personal diary indicates my experience; including a fear of becoming "hooked." I had not at that time read Howard Becker's (1963) studies: "Becoming a Marijuana User" and "Marijuana Use and Social Control", however, from conversations with people I had a good idea what they were about.

And in October 1984 my personal diary expressed my fears of the time:

Becoming a	Marijuana User
	Ethnomethodologist

"Becoming an ethnomethodology user"
Hello me. I am now on page 112 of Benson, D. and Hughes, J.A. "The Perspective of Ethnomethodology." I have also read recently Ch. 3 of Garfinkel (1967) re. the documentary method.

I feel suddenly a doubt. To be or not to be an ethnomethodologist, a slave, it seems: to throw my all in, under one banner, is slavery to a perspective. It is to enter a community of "Cobelievers"; they smile their evangelical smiles of 'rightness' of 'right-mindedness.' They are so 'right'. Benson and Hughes: ethnomethodology is not a replacement for Conventional Sociology, not a replacement for anything, it does not compete, it has a Mona Lisa smirk. It is a critique of Conventional Sociology. But essentially ethnomethodology is its own person.

My doubt is that by taking 'ad-hoc practices' as subject matter, the subsequent accounting for such 'ad hoc practices' is similarly attained by ad-hoc practices by writer and reader . . .

Infinite regress, redress. I find it absorbing. But what of it? Where does it lead?

Do I want to *be* an ethnomethodologist? Being. I am. Join a community of co-believers. Why? Win approval, say the right things, be awarded the Ph.D., succeed in my own Odyssey.

Every so often I question my own motivations. My own life-projects in the life-world: my retrospective—prospective—creative—interpretation of my own being-in-the-life-world. . .

* * *

Actually I do 'believe' I do share the ethnomethodological point of view a la Benson and Hughes. . . . It is a perspective which enables me to 'make sense' of the world. Is it supposed to fulfill that function? It is a 'way of seeing' which I enjoy and find meaningful and *unalienating*. Whereas all the theories derived from coded and quantified obseervables leave me cold: alienated, as a matter of existential fact . . . (Diary Note: Sun. 28/Oct./1984).

Before writing the present autobiographical chapter of how I gradually engaged with ethnomethodology, I first re-read much of my old personal diary to recapture the mood, finding it extraordinarily embarrassing, but fascinating in an eerie way.

[Since writing all those things I have split up with the girlfriend, completed my field research, begun writing up and am currently employed by Lancaster University as a lecturer working very happily on a research project with John Burgoyne.]

Also, before writing this chapter, and after re-reading my diary, I have read Becker (1963): "Becoming a Marijuana User" and "Marijuana Use and Social Control." I am struck by both of these accounts, so closely do they match my experience of "becoming an ethnomethodology user."

Becker (1963, pp. 42-43) on "Becoming a Marijuana User":

What we are trying to understand here is the sequence of changes in attitude and experience which lead to *the use of marijuana for pleasure*.

He outlines three steps in this sequence:

No one becomes a user without (1) learning to smoke the drug in a way which will produce real effects;	"Learning the Technique" (p. 46)
(2) learning to recognize the effects and connect them with drug use (learning, in other words, to get high);	"Learning to Perceive the Effects" (p. 53)
and (3) learning to enjoy the sensations he perceives (Becker, 1963, p. 88).	"Learning to Enjoy the Effects" (p. 53)

Similarly I had to: (1) learn the technique, that is, to ignore the practical motives in any interaction and focus analytically on its organization (compare Sharrock and Watson 1984 vs. Bruce and Wallis 1983); (2) learn to perceive the effects, that is, to maintain the orientation long enough to witness what it is ethnomethodologists are on about; and (3) learn to enjoy or appreciate the effects, that is, to see that ethnomethodology could be an attractive alternative to conventional sociology:

1. *Learning the Technique*

> The first step in the sequence of events that must occur if the person is to become a user is that he must learn to use the proper smoking technique so that his use of the drug will produce effects in terms of which his conception of it can change.
>
> Such a change is, as might be expected, a result of the individual's participation in groups in which marijuana is used. In them the individual learns the proper way to smoke the drug. This may occur through direct teaching (Becker 1963, p. 47).

For me I would read Garfinkel or Sacks for a while and get the feeling that I was missing something. So Barbara and I would discuss it and somehow I would see it more clearly. Next time I read the same pieces it would make more sense.

> Many users are ashamed to admit ignorance and, pretending to know already, must learn through the more indirect means of observation and imitation (Becker 1963, p. 48).

Barbara and I went along to a seminar she had organized. Various sociologists and practitioners of ethnomethodology would be there. I was very nervous in case they found me out, but I decided to play cool. We chatted casuallay over coffee in the senior common room. Before long someone else turned up, and somebody suggested shall we go find a room.

We found one and one of them took out some stuff he had acquired in town. He had been hanging around an advertising agency one day a week and had picked up some sketches, designs, odds and ends scraps of paper with "kiswali" on it ("kiswali" is a kind of gobbledegook used on advertisement designs to indicate where text might go and how it might look in relation to the pictures or whatever). They began analyzing it *ethnomethodologically:* observing the ways in which a sheet of drawings—a mock-up advert—was organized. Various people made suggestions. Someone raised an eyebrow at me. I did not panic. I just played cool and also made a suggestion. Most of them ignored it and carried on their own discussion.

On this occasion I could not see the point. On later occasions, however, I could.

2. *Learning to Perceive the Effects*

> It is not enough . . . that the effects be present alone, they do not automatically provide the experience of being high. The user must be able to point them out to himself and consciously connect them with having smoked marijuana before he can have this experience. Otherwise, no matter what actual effects are produced, he considers that the drug has had no effect on him. Such persons believe the whole thing is an illusion and that the wish to be high leads the user to deceive himself into believing that something is happening when, in fact, nothing is (Becker 1963, p. 49).

At first I could not see the difference between the analytic motive of ethnomethodology and the pragmatic motive of every day life. I thought that the analytic attitude was potentially useful in enabling one to see in finer detail how social reality was constructed hence how it could be improved. My understanding of Garfinkel's breaching experiments was something like: if you can understand the rules by which people operate—by use of the analytic attitude then there will be fewer misunderstandings.

I strove to look *just* analytically, without the concern for working out ways to improve things. It took me many attempts at reading various texts to see it. Eventually through reading Ryave and Schenkein's (1974) "Notes on the Art of Walking" I learned to perceive the difference between the pragmatic attitude and the analytic attitude. But this perception, while it felt like an achievement at first, then left me worrying where was the point. As I said in my diary:

> Anyway, having recognized the fascination of the study [Ryave's and Schenkein's], the heightening of one's awareness of the everyday quotidian, mundaneity of walking and related activities, I am left wondering: what is the point . . .
>
> I.e., the *social purpose* of ethnomethodology (apart from its pure analytic interest) is what I cannot see . . .
>
> It seems to me that the 'objectivity' of positivism is replaced in Ethnomethodology by a kind of self-neutered neutrality, of questionable 'interest.' A sort of denial of the *social purpose* of sociology . . . (Diary Notes: Tues. 30/Oct./84).

I had managed at last to perceive the difference between a pragmatic perspective and a purely analytic, non-pragmatically motivated, perspective on any social interaction whatsoever, even as routine as walking down the street. However, I was not sure I liked how it felt, being able to achieve this "analytic effect."

Through reading Ryave and Schenkein's paper I definitely experienced something different. I had always noticed the niceties in the navigational art of walking down a street so as to arrive at one's destination without actually bumping into someone or something. However, through a reading of their article I felt something new: an analytic attitude, cut off from any concern to avoid bumping, toward the organization of the activity walking in the street. It was the recognition *that* even walking is such a massively organized affair. Suddenly I could see the intricate web of expectations and technique which participant walkers made routine use of in almost complete ignorance of, that is, in absolutely taken-for-granted ways. It felt like an entirely new vista on every day phenomena had opened. From now on, provided I could remember how to do it, I could switch into this detached non-practical attitude. Although every new occasion of doing so would be different and success in achieving "the attitude" and its effects not guaranteed.

3. *Learning to Enjoy the Effects*

> One more step is necessary if the user who has now learned to get high is to continue its use. He must learn to enjoy the effects he has just learned to experience. Marijuana-produced sensations are not automatically or necessarily pleasurable. The task for such experience is a socially acquired one, not different in kind from acquired tastes for oysters or dry martinis. The user feels dizzy, thirsty; his scalp tingles; he mis-judges time and distances. Are these things pleasurable? He isn't sure. If he is to continue marijuana use he must decide that they are. Otherwise, getting high, while a real enough experience will be an unpleasant one he would rather avoid.
>
> In no case will use continue without a redefinition of the effects as enjoyable. This redefinition occurs, typically, in interaction with more experienced users who in a number of ways, teach the novice to find pleasure in this experience which is at first so frightening" (Becker 1963, pp. 53-54).

Thus, in my case, my supervisor was able to reassure me when I panicked unable to see the point of being so analytically detached. This she did by describing her first encounter with the ethnomethodological perspective, which seemed to cause an even more unnerving reaction in her. She said that, suddenly everything around her seemed as if it could fall apart:

> You worry in case you are the only one who is holding it all together.

This was the effect of the insight achieved by adopting the analytic attitude.

We could discuss it in terms of our shared knowledge of Carlos Casteneda's work, and suddenly for me too it was liberating: to let go of any need to hold on to the practical purpose or point to it all. That is not to say that on other occasions I did not have a few "bad trips"; that is, from time to time I would become anxious that again I had lost the point, although usually through conversations with my supervisor I could let go of the practical point and seize the analytic one if I wanted to.

Nonetheless as Becker notes:

> Learning to enjoy marijuana use is a necessary but not sufficient condition for a person to develop a stable pattern of drug use. He still has to contend with the powerful forces of social control that make the act seem inexpedient, immoral or both . . .
>
> The career of the marijuana user may be divided into three stages, each representing a distinct shift in his relation to the social controls of the larger society and to those of the subculture in which marijuana use is found (pp. 59-61).

In the first stage one is a "beginner," in the second one is an "occasional user" and in the third one is a "regular user." To enter each stage requires that one surmounts three different kinds of social control: (1) to become a beginner one must surmount the problem of locating a supply; (2) to become an occasional user one must solve the problem of how to do so secretly; and (3) to become a regular user one must overcome the difficulty in seeing regular usage as a moral activity.

How did I cope with these social controls?

1. *Becoming a Beginner: Supply*

The supply was easy although I concede it would not be for everyone. In the first place I had a contact who could provide references; this enabled me to know where to go. Admittedly, the stuff was not lying around the business school library but the main university library had several good books and articles.

> In order for a person to begin marijuana use, he must begin participation in some group through which these sources of supply become available to him, ordinarily a group organized around values and activities opposing those of the larger conventional society (Becker 1963, p. 62).

Thus, Barbara introduced me to a social set, different from the usual business school crowd, which included ethnomethodologists, conversation analysts and assorted ethnographers and eventually a whole bunch of sociologists at the 1985 B.S.A. Summer School. However, my participation in such circles was only partial.

2. *Becoming an Occasional User: Secrecy*

Because I spent most of my time outside such circles where ethnomethodology use was a part of social intercourse, I had to learn how to practice it on my own while in normal "non-deviant" settings. Thus, when taking field-notes I learned to focus on the classroom's organization, the organization of talk, nods, winks, smiles, raised eyebrows, questions asked and more. All of which I could do uninterruptedly as the lecture or class proceeded. It became awkward when, at coffee breaks, for instance, members asked such things as: what do you find to write about? Then I felt embarrassed; I had hoped my note-taking behavior would not have been perceived to be too abnormal. So sometimes I would not take as many notes as I could, *in class,* but would try to write them up later. This was more acceptable to the members but it meant that I did not trap on paper as many insights as I experienced while sitting there:

> Marijuana use is limited to the extent that individuals find it inexpedient or believe that they will find it so. This inexpediency, real or presumed, arises from the fact or belief that if non-users discover that one uses the drug, sanctions of some important kind will be applied. . . Although the user does not know what specifically to expect in the way of punishments, the outline is clear: he fears repudiation by people whose respect and acceptance he requires practically and emotionally. That is, he expects that his relationships with non-users will be disturbed and disrupted if they should find out, and limits and controls his behavior to the degree that relationships with outsiders are important to him (Becker 1963, pp. 66-67).

I found, under the analytic attitude of ethnomethodology, much more going on in any particular classroom session than, for example, when I was under the attitude of everyday life. I was aware that under the analytic attitude, my note-taking in front of part-time students might appear abnormally frantic, detailed, and elaborate. This in turn might occasion sanctions. Indeed it did from time to time; once a younger member of the part-time program sitting next to me seized my notes forcibly to see what I had written. This felt awful. I knew he would not understand the reason for making such close observations. He called out something like: "He writes down *everything!* . . . every little detail!" And he was about to elaborate with examples from my notes such as "So and so—late again can't find a chair"— when I grabbed the notes back.

Such fears of being found out restricted my pattern of ethnomethodology use in field-note taking at first.

> Regular use . . . is a mode of use which depends on another kind of attitude toward the possibility of non-users finding out, the attitude that marijuana use can be carried on under the noses of non-users, or, alternatively, on the living of a pattern of social participation which reduces contacts with non-users almost to zero point. Without this adjustment in attitude, participation or both, the user is forced to remain at the level of occasional use. These adjustments take place in terms of two categories of risks involved: first, that non-users will discover marijuana is one's possession and, second, that one will be unable to hide the effects of the drug when he is high with non-users (Becker 1963, p. 68).

Clearly, the first risk did not apply—no one could find ethnomethodology *on* me—but the second risk did apply. My frantic, highly detailed note-taking could betray my use of the ethnomethodological perspective, if read.

3. *Becoming a Regular User: Morality*

> Conventional notions of morality are another means through which marijuana use is controlled. The basic moral imperatives which operate here are those which require the individual to be responsible for his own welfare, and to control his behavior rationally. The stereotype of the dope fiend portrays a person who violates these imperatives (Becker 1963, pp. 72-73).

From various things my supevisor said and from various things I read, I felt that once I had learned the technique of seeing the world through the analytic ethnomethodological attitude then there would be no way back to normal. My interest in normal science, "conventional sociology" (Benson and Hughes 1983) would be gone, no longer would I be able to see the point in developing theories of causation or explanation, no longer would I be able to cherish hopes of writing books which offered solutions to the problems of management learning. No longer would I wish to earn an honest living selling my consultancy skills like any other normal member of the business school fraternity.

In short, I was afraid ethnomethodology would take over my life. The intellectual hedonism of ethnomethodology use would prevent me functioning, as a normal upholder of conventional mores about the proper way of doing management research, in order to improve things like any other "social engineer."

I was impressed by the last few pages of Benson and Hughes (1983):

> Social engineers, to put it this way, are already there in the society as its members, continually producing it. So, to push the image a little further the ethnomethodologist, as a social scientist, is not an engineer among mechanical idiots but another engineer who happens to be interested in the practices of engineering.
>
> Where might this interest lead? To better engineering practices? More humane ones? Or simply to the creation of a useless lot of narcissistic engineers collected together in a mutual admiration society? . . . Garfinkel expressed the relationship between ethnomethodology and sociology and what, respectively, they might promise. He made the comparison between chemistry and alchemy. Alchemy promised the world; a promise to be redeemed by the transmutation of base metals into gold and the discovery of the elixer of youth. Chemistry's aims were more modest. Although they used much the same painstakingly developed technology of alchemy, these were put to different ends. Different sets of questions were asked about the world, and different theoretical schemes developed. For many years the chemists did not have it all their own way. Alchemists kept pointing out that they were getting nearer all the time to the final goal of transmuting lead into gold . . . In this kind of climate, the early chemists had very little chance, especially when they occasionally remarked that they were not too much interested in making gold but much more fascinated by what happened when a glass was placed over a lighted candle. No doubt as time went on some of the more generous-hearted alchemists began to admit that perhaps chemistry did have a part to play in alchemy, though only of some limited 'micro' relevance (Benson and Hughes 1983, pp. 199-200).

Obviously with hindsight, chemistry, which began as a non-useful interest with no promise of solving anything worthwhile, eventually proved more worthwhile than the pipe-dreams of conventional alchemy. Perhaps it is too early to say what will happen vis-à-vis ethnomethodology and sociology. As Benson and Hughes (1983, p. 200) say:

> . . . the challenge for conventional sociology remains: to stay committed to common-sense problems and rest content with analyses that can never be more than yet another folk version of the world; or to subject the everyday, mundane, routine world to analysis in an effort to describe how common-sense understanding themselves are generated. The two are not compatible.

Consequently, for me it seemed like an "either/or." And I could sense the moral disapproval that business school academics would pour on one who stopped trying to solve management problems on the grand scale and simply sought to describe how common sense understandings are themselves generated, that is, *managed* accomplishments.

A further doubt was that I could not see how even descriptions of ethnomethodology's, admittedly different, phenomena could likewise ever be "more than yet another folk version of the world." My only solution was to "be aware" of this problem realizing that "folk versions" were fine; whether of conventional sociology's chosen phenomena (see Hammersley 1985) or of ethnomethodology's chosen phenomena.

I did not wish to become a slave to the ethnomethodological perspective as my diary note at the start of this chapter indicates. However, in order to use this form of analysis I had to convince myself that I could always change the phenomena I looked at if I wished to. I need not *just* focus on how common sense understandings are generated. I overcame the fear that once an ethnomethodologist always an ethnomethodologist: I could control it.

> In short, a person will feel free to use marijuana to the degree that he comes to regard conventional conceptions of it as the uninformed views of outsiders and replaces those conceptions with the "inside" view he has acquired through his experience with the drug in the company of other users (Becker 1963, p. 78).

So far this chapter has been an autobiographical account, perhaps rather fancifully compared to becoming a regular marijuana user, of how I came to use the ethnomethodological analytic attitude in my analysis of data; which inevitably infused my data-collection approach, focuses and practices. However, three further points need to be made.

First, ethnomethodology is not being downgraded by this metaphorical comparison to marijuana. My point is that ethnomethodology has, to a considerable extent, stood outside conventional sociology as a separate, almost "deviant" minority interest and perspective. Its practitioners and exponents have often claimed it is not compatible with conventional sociology's problematic. As Benson and Hughes (1983) outlined the conflict: one may carry on producing more and more sophisticated "folk versions" of the social world and the problem of social order, as conventional sociology does, or one may take the orderings of these "everyday folk versions" as the phenomena to be analyzed. In both cases one cannot escape using common sense practical reasoning, but one may choose which problem one wishes to address. Conventional sociology seeks to address the problem of order as if it were "out there" waiting to be explained by theorizing and empirical research. Ethnomethodology seeks to address the problem of order that it finds *in* any account of social structures; indeed it starts from the view that it is through members' account-making practices that the social structures "out there" are constituted.

The dispute between conventional sociology and ethnomethodology appears to turn on where the problem of social order is to be located: whether it is "out there" amidst members' power relations, or whether it is "within" members' methods and practices of recognizing and producing their power relations. Anybody examining any piece of data whatsoever chooses between these perspectives. The conventional sociologist, like anybody-else, in everyday-life is concerned, as Garfinkel said, to:

> . . . elect among alternative courses of interpretation and inquiry to the end of deciding matters of fact, hypothesis, conjecture, fancy and the rest. . . (Garfinkel 1967, p. 77).

And, of course, as a field researcher I was concerned to do this as I participated along with the part-time M.B.A. students, in my field research. However, what I also chose to do was to analyze members' practices for electing among alternative courses of interpretation. I did not do this exhaustively at every elective juncture, because such junctures are infinite (see Garfinkel 1967, pp. 24-31), rather I did this illustratively and selectively. That is, I controlled my use of ethnomethodological perspective, adopting it when I thought it would offer interesting insights into members' social constitution of their experience (and I include myself among the members), since the purpose of my thesis was to illuminate the experience of part-time masters' students of business. My use of the ethnomethodological perspective was not constant, but it was more than occasional. I used it regularly in the course of my conventional ethnographic field research, whenever I thought it would be illuminating.

Second, ethnomethodology, like marijuana, can be used to produce certain illuminating effects. And as with marijuana, learning to use ethnomethodology is rather like learning a mystique. One does it by socially interacting with and learning from others who use it regularly. But to adopt a pattern of regular use one must, as Becker (1963, p. 68) notes, either take up the attitude that one can use it under non-users' noses without being found out, or live a pattern of social participation which reduces contacts with non-users almost to zero-point. Living and working within the business school as I did in the course of my field research, I could not adopt the latter option. This left me on the horns of a dilemma as a practical field-reseracher within the business school since everyone there—apart from my supervisor—was committed to the perspective of everyday life that is shared by conventional sociology (as far as I could tell).

My mode of coping was, therefore, to use the ethnomethodological perspective either under members' noses, as I did at times when taking field-notes, or "in secret", as it were, when I was in my room analyzing the data, or with my supervisor discussing the data.

Third, having done my field research and written my substantive chapters, I would suggest that the use of ethnomethodology should not be seen as a kind of morally deviant activity within conventional sociology, for the use of the ethnomethodological perspective enriches the insights gained in the process of doing ethnography. It allows one to go beyond simple description of what the members saw and did and felt; to analyze how they put their "persuaded versions" of the facts, their feelings and experience together. It allows one to unpick the taken-for-granted ways in which they *elected,* often without noticing it, among alternative courses of interpretation in specific settings. The part-time M.B.A. students decided questions of fact, conjecture, justice and more in their everyday interaction and out of such "interpretive work", their experience of the course and the business school was constituted. Analyzing this from the ethnomethodological perspective allows one to, as it were, "catch the work of 'fact production' in flight" (Garfinkel 1967, p. 79).

Thus I would suggest, that while people, including conventional sociologists pursuing their problematic, in everyday-life are concerned to inquire into the nature of social reality with a view to greater understanding and often with the aim of becoming better at social engineering, ethnomethodologists can offer insights into the ways in which members' very methods of managing inquiry accomplishes what they find. And in the case of part-time M.B.A. students inquiring into management, via formal management education, the ethnomethodological perspective illuminates how they managed to accomplish what they learned.

ACKNOWLEDGMENT

The research was conducted as the basis of my doctoral thesis (Fox 1987) and was supported by a grant from the Economic and Social Research Council.

REFERENCES

Becker, H.S.
1963 *Outsiders: Studies in the Sociology of Deviance.* New York: Free Press.

Becker, H.S. et al.
1961 *Boys in White: Student Culture in Medical School.* Chicago: University of Chicago Press.

Benson, D. and Hughes, J.A.
1983 *The Perspective of Ethnomethodology.* London: Longman.

Bittner, E.
1974 "The Concept of Organisation." In R. Turner (ed.), *Ethnomethodology: Selected Readings.* Harmondsworth: Penguin.

Bruce, S. and Wallis, R.
1983 "Rescuing Motives." *The British Journal of Sociology* 34(1):61-71.

Burgess, R.G.
1984 *In the Field: An Introduction to Field Research.* London: George Allen and Unwin.

Fox, S.
1987 *Self Knowledge and Personal Change: The Reported Experience of Managers in Part-Time Management Education.* Unpublished Ph.D. thesis, University of Manchester.

Garfinkel, H.
1967 *Studies in Ethnomethodology.* Englewood Cliffs, NJ: Prentice-Hall.

Gold, R. L.
1958 "Roles in Sociological Fieldwork." *Social Forces* 36:217-223.

Hammersley, M. and Atkinson, P.
1983 *Ethnography Principles in Practice.* London: Tavistock.

Junker, B.
1960 *Field Work.* Chicago: University of Chicago Press.
Ryave, A. L. and Schenkein, J. N.
1974 "Notes on the Art of Walking." Pp. 265-274 in R. Turner (ed.), *Ethnomethodology: Selected Readings.* Harmondsworth: Penguin.
Sacks, H.
1974 "On the Analysability of Stories by Children." Pp. 216-232 in R. Turner (ed.), *Ethnomethodology: Selected Readings.* Harmondsworth: Penguin.
Sharrock, W. W. and Watson, D. R.
1984 "What's the Point of 'Rescuing Motives'?" *The British Journal of Sociology* XXXV(3):435-451.
Silverman, D.
1978 "'K is Mentally Ill' The Anatomy of a Factual Account." *Sociology* 12(1):23-53.
Thomas, A. B.
1980 "Management and Education: Rationalization and Reproduction in British Business?" *International Studies of Management and Organization* X(1-2):71-109.
Whitley, R., Thomas, A.B., and Marceau, J.
1981 *Masters of Business: The Making of a New Elite?* London: Tavistock.
Zimmerman, D. H. and Pollner, M.
1971 "The Everyday World as a Phenomenon." In Douglas (ed.) *Understanding Everyday Life: Toward the Reconstruction of Sociological Knowledge.* London: Routledge and Kegan Paul.

DECISION TAKING IN THE FIELDWORK PROCESS:

THEORETICAL SAMPLING AND COLLABORATIVE WORKING

Janet Finch and Jennifer Mason

One of the key ways in which qualitative or fieldwork methods differ from social surveys is in the sampling or selection of the people and situations that are studied. In surveys, such decisions are made once-and-for-all at the beginning of a project, and follow formalized statistical procedures for sampling. In fieldwork, such decisions are taken at various stages during the course of the project on the basis of contextual information. To outsiders who are not privy to the changing contextual basis of this project, research decisions can look rather ad hoc.

In this chapter we are going to discuss the questions of whether and how decisions about fieldwork sampling can be taken in a systematic way. We will use our experiences of working together on the research study of Family

Studies in Qualitative Methodology, Volume 2, pages 25-50.

ISBN: 1-55938-023-3

Obligations to provide the contextual information necessary for a discussion of systematic decision taking.[1] At the time of writing we are at the mid-point of this study with most of the data collected but much of the formal analysis still to be done, so we cannot relate to our sampling strategies to our final analysis. However, what we have tried to do is to reflect the blend of practical and intellectual considerations that form the basis of decision taking in the fieldwork process, at a stage when these are fresh in our minds.

We begin by briefly describing the Family Obligations project before outlining the principle of theoretical sampling, which represents a possible model of systematic selection in qualitative research. In the main part of this paper we discuss how we attempted to develop and apply this principle in our own research and we use our experience to suggest more generally the problems and possibilities of this approach. A further important theme in our discussion is the extent to which working collaboratively facilitates systematic qualitative research.

THE FAMILY OBLIGATIONS PROJECT

Our project involves investigating patterns of support, aid and assistance, of both practical and material kinds, between adult kin in a survey population based in the Greater Manchester area. As well as being interested in patterns of support, we are exploring concepts of obligation and responsibility to assist one's relatives, the circumstances in which these come into play, and the processes through which these are related to actions. We have used a conceptual framework for studying family obligations based on a contrast between two different ways of conceptualizing obligations: as moral norms and as negotiated commitments (Finch, 1987). On the one hand, family obligations might be seen as part of a structure of normative rules that operate within a particular society, and which simply get applied in appropriate situations. On the other hand, they might be seen as agreements that operate between specific individuals and are arrived at through a process of negotiation. This negotiation may be explicit, or more likely may be covert. We would argue that a full understanding of what family obligations mean and how they operate almost certainly contains elements of both of these.

Our perspective suggests that these norms are not really like rigid and precise rules that must be followed, or that are imposed on passive individuals. Rather they are general guidelines for "proper" or "correct" behavior toward one's relatives, which "everyone is aware of," but that need to be interpreted or tailored in specific situations. It is at this level of interpretation that the notion of negotiated commitments comes to the fore: if it is the case that norms are not sufficiently detailed or universally applicable to be used straightforwardly in concrete situations among relatives, then in what ways are actual commitments and responsibilities negotiated? This part of our perspective casts "negotiation" in broad terms, allowing not only for "round the table" explicit negotiations, but also for other processes by which particular patterns of obligation become "obvious" to members of kin groups.

We are not going to discuss the substantive issues in the project here, but it is necessary to give some detail so that we can explain the nature of the decisions that we had to take about selection and sampling. The perspectives upon the study of family obligations that we have outlined here guided the overall planning of the project, and led to a research design that included both a large-scale quantitative survey and also a second stage of qualitative fieldwork, mainly based on in-depth interviews. In the survey we were concentrating solely upon data about normative beliefs, and at the second stage we wanted to use qualitative techniques to understand more about the complexity of beliefs, as well as the relationship between beliefs and actions, and how people actually negotiate commitments with their own relatives.

It is this second stage that we discuss in this chapter, as this was where we were trying to put theoretical sampling into practice. We planned to use the survey population as a sampling frame from which to select individuals for more detailed study. We had a total survey population of 978 randomly selected individuals over the age of eighteen, of whom 85% had agreed at the time of the survey that they would be willing to be reinterviewed.[2]

We, therefore, had the possibility either of choosing another randomly selected group for more detailed study or of targeting particular subgroups. A further consideration was that we hoped to be able, in some cases, not only to reinterview a survey respondent, but also to interview members of his or her kin group, thus building up a more complete picture of negotiations within families. In a sense, therefore, we were operating two levels of selection: interviewees from the survey population, then the kin groups of some of those interviewees. There was a limitation of numbers in that we had budgeted for 120 interviews at this second stage, but other than that we were free to be guided by theoretical considerations, and by our preliminary analysis of the survey data, in deciding whom to interview. However, we needed a strategy that was very flexible so that we could: (1) change direction as we went along if necessary, (2) leave open the possibility of doing more than one interview with some respondents, and (3) maintain the possibility of interviewing other members of the kin group.

We had, therefore, in effect set ourselves the task of being both flexible and systematic in our selection of interviewees for the qualitative stage. In practice this seems commonly to be the aim of much fieldwork based research, and certainly sits comfortably with a strategy of theoretical sampling.

A GUIDING PRINCIPLE: THEORETICAL SAMPLING

What are the main themes in the existing literature on fieldwork that can help to guide our thinking on this process? In addition to the emphasis upon selection as an on-going process, the two important themes seem to be first the interplay of theory and data, and second that the analysis of data is a process that continues throughout the project rather than occurring as a discrete phase after the data collection is complete. Theory should guide data collection and the on-going analysis of data should feed back into theory, which in turn guides the next phase of data collection. Most fieldwork researchers would acknowledge that this is the model to which they aspire, but as other commentators have noted this process is often not put into practice very effectively, leaving (and there are often) quite serious gaps between theory and data (Hammersley and Atkinson 1983, p. 174).

The concept of theoretical sampling is probably the most common way of translating this model of the research process into guidelines about selection of research situations or informants. Certainly this is the concept that guided our own thinking and it forms a central focus of this chapter. We offer an account of our own experience of trying to put into practice the notion of theoretical sampling, and use this to draw out some general principles about how qualitative research can be done in a systematic way.

The term "theoretical sampling" is generally associated with Glaser and Strauss's treatise on the discovery of grounded theory (1967), but its logic and practice has become part of a tradition of qualitative research (Bertaux 1981; Schwarz and Jacobs 1979; Baldamus 1972; Hammersley and Atkinson 1983). Essentially, theoretical sampling means selecting a study population on theoretical rather than, say, statistical grounds. The underlying logic is one of analytical rather than enumerative induction which were distinguished many years ago by Znaniecki in the following way:

> Enumerative induction abstracts by generalisation, whereas analytic induction generalises by abstracting. The former looks in many cases for characters that are similar and abstracts them conceptually because of their generality, presuming that they must be essential to each particular case; the latter abstracts from the given concrete case characteristics that are essential to it and generalises them, presuming that insofar as they are essential, they must be similar in many cases (Znaniecki 1934, pp. 250-251).

This means that theoretical sampling involves a search for validity of findings, rather than representativeness of study population. However, for some degree of generalization to be made about the consequent research findings, it is vital that the processes of theoretical sampling (as well as data presentation) be *systematically* carried through and documented.

Yet when viewed apart from its context, certain aspects of theoretical sampling can appear ad hoc and unsystematic. In particular, from a positivist standpoint, where research decisions are made in advance of operationalization and tested through hypotheses on a randomly sampled population, the continual making of decisions throughout the course of the research can itself appear very unsystematic. Of course this is partly because the two endeavors are not entirely comparable: qualitative researchers following theoretical sampling are generally looking to build theory from data rather than to test hypotheses on representative populations. At the same time, however, we are not suggesting that qualitative researchers have license to be unsystematic in their decision making simply because it cannot be done in "one go." Rather, we would emphasize that the validity of the qualitative researcher's interpretations depends in part upon the quality and relevance of their in-process decisions.

But what does this actually mean in practice? How is theoretical sampling *done?* Given that such importance is placed upon in-progress decision making in particular research settings, then it is inappropriate to set down in advance a series of general rules about how to make informed and systematic decisions. Instead we describe below what we did, and on what basis we made our decisions. As well as giving a more situated feel for theoretical sampling, this will in a sense provide the data from which to extrapolate some more general observations at the end.

A GUIDING PRACTICE: COLLABORATIVE WORKING

A vital part of the way in which we have put theoretical sampling into practice in the Family Obligations project has been through collaborative working. Neither of us had worked in a close collaboration of this kind before, and, therefore, part of our task was to develop ways of working together effectively. Janet Finch had set up the project and organized the survey fieldwork[3] and Jennifer Mason joined her at the stage when that was completed. It had been agreed that we each would take an equal share in planning and conducting the qualitative fieldwork, but we wanted to ensure that our contributions were fully integrated with each others' and that the fieldwork would be a genuine collaboration. We tried to achieve this in a number of ways.

Joint Discussions and Planning

From the beginning we held regular joint discussions about our plans, strategies and practice. Our early discussions centered on ways of working together, how often we should meet and so on. We decided at that time always to take decisions together about overall strategy and practice. This may seem a rather obvious point to make about collaboration, but we know from our own contacts with other researchers, as well as from a few published accounts, that decisions are not always taken openly and explicitly in research teams, leaving the opportunity for misunderstandings and disagreements to arise about what strategy is actually being followed (Platt 1976; Bell 1977; Porter 1984).

Division of Labor

One aspect of our practice that we have always discussed, rather than assumed, is our division of labor. Jennifer Mason was employed for three years full-time on the Family Obligations project, and Janet Finch, who holds a full-time university teaching post, had arranged to have one year seconded to the research project. This meant that our first year of working together could involve collaboration on a full-time basis for both of us. One of the reasons Janet Finch had set up the project timing and staffing in this way was so that she could maintain a full involvement in the fieldwork stage, and the research process that we discuss in this chapter all took place during that time.

Therefore, when discussing our fieldwork division of labor we could take an equal share of the tasks. As a result, we were able to organize ourselves to conduct half of the interviews each, as well as to structure in time to keep up to date with what the other was doing. Some of the procedures we used to maintain an ongoing preliminary analysis of the fieldwork are described later.

Collective Research Diary

One of the mechanisms that we used to achieve this was a collective research diary. We agreed at an early stage that we would record all substantial discussions that we had about the research, and would always give each other a copy of any notes that were made individually. In each of our joint meetings, we would agree that one of us would take notes and produce a record of the meeting, so that our collective research diary is made up of contributions from both of us.

It is this collective research diary that forms the basis of our discussion here. We have decided to include extracts from this diary and we have kept these in their raw state with no editing, so that readers can see the actual process of decision taking at work. When these notes were written, we had no idea that we might publish them in this way, and they were written solely for our own use although we did anticipate reflecting on them and using them in our analysis. We have little idea whether our procedures match other people's since personal research diaries of this kind are seldom made publicly available. Equally, we know little about the actual day-to-day procedures that people use to work collaboratively, and how far ours are distinctive. We have decided to include our raw notes as extracts because of the lack of discussion in the literature about these issues, although the importance of making public the research process in this way has been noted by others (Stenhouse 1980; Burgess 1984).

INITIAL SELECTION OF INTERVIEWEES

We shall deal with issues of selection as they occurred chronologically telling our story in the order in which things actually happened, but also drawing out theoretical points as we go along. Figure 1 provides a "map" of the sequencing of events in the story.

The first set of decisions that we had to make concerned which people in the survey population would be chosen for more detailed study. We also had to decide how many to select. There were a number of interwoven considerations here concerning: principles of selection; the possibility of having subgroups and how many of these we would realistically include; strategies for selection to retain maximum flexibility especially to accommodate successful attempts to move outward to some interviewees' kin. Extract A is taken from the record of the planning meeting in which we talked through all these issues.

Extract A: Notes from the Record of Planning Meeting 5/8/86 (recorded by Janet Finch)

Our final decision was as follows:

We will begin selecting interviewees from two sub groups: people who have been divorced and/or remarried; young adults (under 25 at the time of the survey).

We will begin with women, until we have our selection procedures running smoothly and have decided on the merits of employing a male interviewer.

We will begin with five from each group, randomly selected from our list of people who meet these criteria. If we have a refusal, we will replace it with another name, selected randomly.

We will build up from there, seeing how far we can get with the kin groups of each, and adding more names from our sample. Our final aim will be to have qualitative data not only from people whose current, recent life experience has involved a renegotiation of family relationships (the basis of the above categories) but also people whose experience is close to the stereotypic norm of family life. If we do not pick up such people via kin groups, we may select from a different sub group of our sample, such as people with large kin networks who were once married or "women in the middle" (of a younger and older generation). We also want our final selection to have a reasonably good social class spread, and therefore we may select interviewees at a later stage which enable us to do that. Social Class I may be a case in point.

This strategy also opens the possibility of finding some other sub group, or individuals with particular characteristics who emerge as important during the course of earlier interviews.

The reasoning which lies behind this strategy included the following considerations:

Figure 1. Decision-Taking Sequence

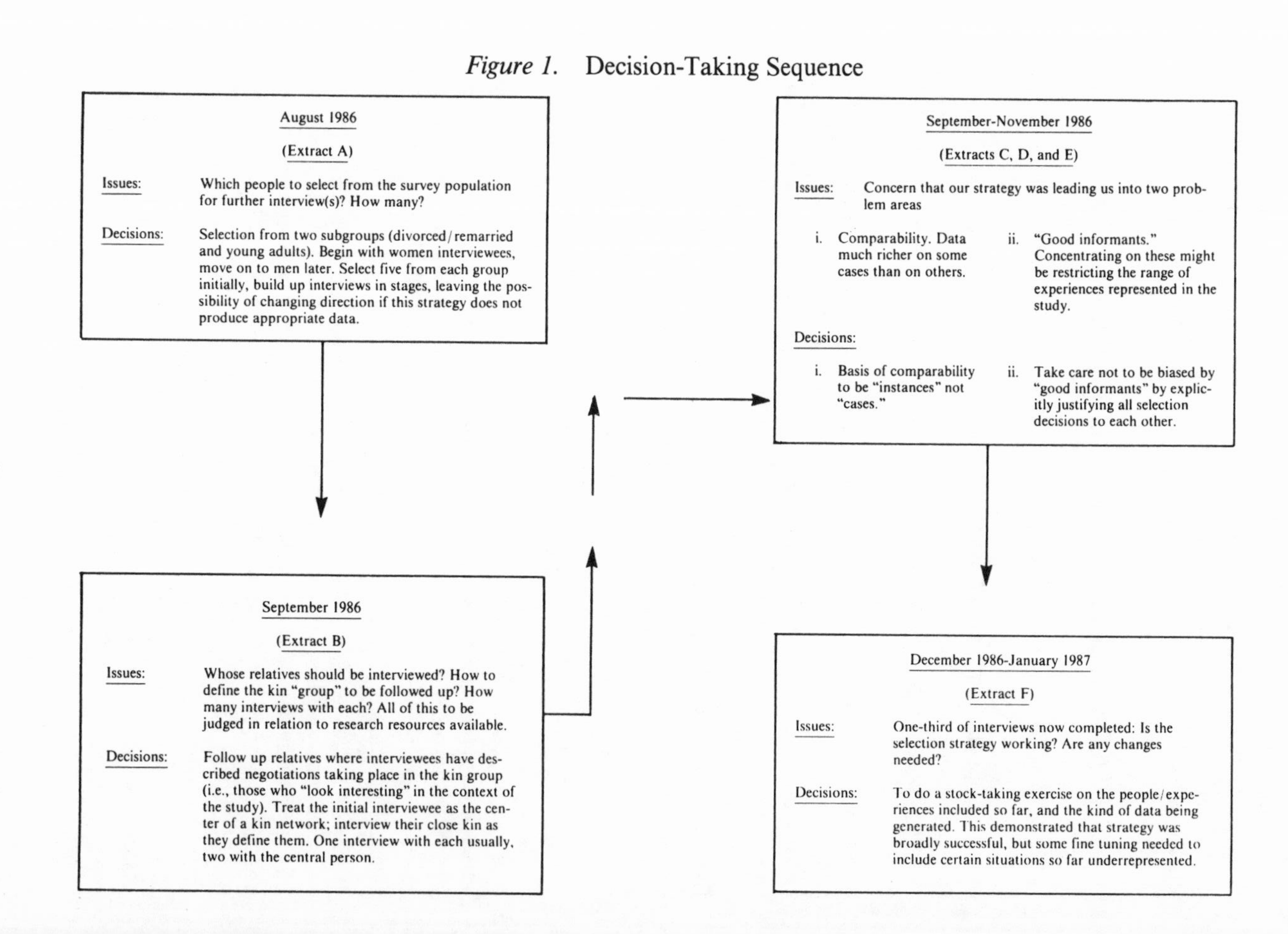

1. A target sample of sixty individuals at the second stage means that four sub groups (i.e., fifteen in each) looks like the absolute maximum. But if we are successful in interviewing kin, then the number selected from the sample will be smaller than sixty. Our strategy of two main sub groups therefore seems realistic.
2. Our two sub groups are selected on the basis of life experience, and we have focused on people where their family relationships are likely to have undergone some renegotiation, which makes issues of obligation more explicit. They are also sub groups which seem to be significant in terms of social change, in that the rules of family obligations are currently probably being written/rewritten—because of divorce/remarriage becoming an increasingly common experience and because of pressures which seem to be creating longer periods of non-independence for young people.
3. We considered principles of selection based upon views expressed in the interview, especially people who gave "standard" answers and people who were in "deviant" minorities. We rejected this because: it seemed less important than the other principles of selection and we didn't want to multiply sub groups; we can probably get a lot out of the interviews themselves in relation to these issues; the second stage interviews are to concentrate upon experience rather than beliefs.
4. We considered adding a third group which would effectively be a control group, composed of people whose experience seems near to the stereotype norm. Although we agreed that it is important to build in the principle of comparison between typical and atypical experience, we decided not to select these initially as a former control group, because: we want to maintain maximum flexibility; to add a third group would mean that we reduce the numbers in the other two at an earlier stage; we may well pick up people whose experience is close to the norm *via* the interviewees initially selected on the other criteria. So we decided instead to treat this as a gap filling exercise at a later stage.
5. Part of the reason for rejecting a formal control group was that this concept derives from an underlying logic which doesn't really fit this stage of the project. We aren't aiming at generalisation based on represenativeness, etc. Instead, we think that our agreed strategy is actually quite faithful to the principles of inductive logic. Indeed, we have left our options open to select cases which emerge as key in the course of our interviews.

It can be seen from this extract that we decided to sample initially from two subgroups: people who had been divorced and/or remarried, and the youngest age group in our population, namely people aged eighteen to twenty-four at the time of the survey. The reasons for this are centrally concerned with theoretical sampling. We quickly rejected the idea of a randomly selected subgroup on the grounds that we were not trying to use our qualitative data to make generalizations based on representativeness (although the survey data of course could be used in that way). Since fieldwork was principally to be concerned with understanding the process of negotiation between relatives, we decided that it would be much more useful to focus upon individuals who might currently or recently have been involved in processes of negotiation and renegotiation of family relationships. We hoped that talking to these individuals would give us access to family situations in which those processes would be most visible. The two groups that we chose seemed to fulfill these criteria.

It is important to underline (as we have found that people sometimes misunderstand what we are saying about our selection strategy) that we did not select these two as comparison groups in the orthodox sense. We were not seeking straightforwardly to compare the experience of young adults and divorced people, but were using both groups as a "way in" to the kind of family situations that we did want to study. We were, for example, very interested in the care of elderly people and how responsibilities for that develop over time within families, but we did not select an elderly subgroup because we hoped that our selection strategy would lead us to such situations in the kin group of the young people and the divorced people whom we had selected. In this sense, we were selecting kin groups (or at least situations in kin groups) as the focus of our study, rather than individual. We recognized, of course, that this strategy might not work, and it can be seen from Extract A that we built in the possibility of reviewing and revising our strategy during the course of the fieldwork.

The major principle that we used to guide our selection, therefore, was theoretical significance: we chose to focus on those groups that would enable us best to evaluate and develop the theoretical ideas and concepts with which we began the project. However, it can be seen from Extract A that we were also juggling with a number of other considerations that helped to shape our overall strategy. We will comment briefly on these.

1. Although we were not aiming at statistical representativeness in this fieldwork study, we did want to ensure that we included a range of experiences of family life. We wanted to include some situations (like divorce) which would be a minority experience and others that were more routine and typical. Further, we could see from our survey data that some subgroups in the population had answered our questionnaire in a distinctive way, and we wanted to ensure that their personal experiences were reflected in our qualitative study. People in Social Class I were one example mentioned in the Extract. There were other groups—for example, people who have experienced unemployment—whose experience we also wanted to include because of its public importance in contemporary Britain. At this initial stage we decided to wait and see if our selection strategy based on two subgroups would lead us to a good range of situations which included all of these. If not, we left open the possibility of rethinking our strategy at a later stage.

2. The inclusion of the experience of non-white people presented us with particular problems since the survey population had only a small number of respondents from ethnic minorities (the refusal rate having apparently been high among these groups). We were also aware that we might want to change slightly the format of our interview to make it appropriate to distinctive cultural experiences, and that we might wish to seek advice on this. At the initial selection stage, while we were strongly committed to not producing an ethnocentric piece of research, we decided that we would come back to the issue of specifically selecting some non-white interviewees at a

later stage in the fieldwork, after we had gotten our procedures operating smoothly.

3. We took a similar decision in respect of interviewing men, although for completely different reasons. We wanted to include both women and men in our study, both in our initial selection of interviewees and as relatives of those selected. However, since we would be conducting interviews in people's own homes, and because we were two female researchers, we were conscious of issues about personal safety and felt that it would be unwise for either of us to go alone into the homes of unknown men (McKee 1983). We considered various possible strategies, including employing a male interviewer, but at this stage in the project we felt dissatisfied with all of the possible solutions. We decided, therefore, that the first group of interviewees selected would be all women, thus buying ourselves time to get our interview procedures running smoothly before we tried to solve the problem of men.

The skeptical reader might accuse us of having put off a great many decisions and conclude that we were unclear about what we wanted to do and unwilling to make firm choices. However, the whole point about trying to achieve an interplay between theory and data, and the logic of analytic induction that underlies fieldwork procedures, is that decisions cannot be taken in a final and irrevocable form before any data are collected. This is because we cannot know in advance of studying some actual cases what are (to put it in Znaniecki's terms) the essentials that we would want to abstract in order to compare them with other cases and to test and refine our generalizations. What we were trying to do at the initial stage of selection was to decide *where to look* for the processes of negotiating obligations in families, rather than to prejudge *what* we would find. In these decisions about where to look we were trying to be guided in a systematic way by theory, while maintaining the flexibility to look elsewhere at a later stage if we had gotten it wrong, and to be able to build in, at a later stage, the comparisons that would emerge as most significant on the basis of our initial cases.

These principles of selection had to then be translated into practical strategy, and we decided that we would proceed by selecting small numbers of people from each subgroup. We actually selected five at a time from each subgroup, giving ourselves the opportunity to assess how many actual interviews we were achieving, and how many relatives we were following up, before going on to select some more initial contacts. We had a total list of 117 young adults who were willing to be interviewed again and 112 people who had been divorced and/or remarried. Each time we selected from these lists we did so using a table of random numbers, since we did not want to develop more detailed criteria about whom to follow up. This, of course, still left open the possibility that at a later stage we could "search" for people who had a particular combination of characteristics that we might wish to include. For example, we could look for people who were unemployed as well as in the young adult group if this experience was not being included in the cases we had initially selected.

At this initial stage we think that our procedures do demonstrate a fairly successful attempt to be both systematic and flexible. We recognize that the actual detail of what we did could not be straightforwardly translated into a different project. We had, for example, a great deal of prior information about potential interviewees based on the survey questionnaires, probably far more than is usually available at the beginning of a piece of fieldwork. But we think that the underlying logical procedures that we were using can certainly be translated to other contexts.

SELECTING KIN GROUPS

Our procedure for sampling individuals from the survey population proceeded as described. After the first five sampled from each group, we continued sampling in small bunches, substituting refusals by another random selection from the same subgroup. Having set the wheels in motion in this way we had to decide whose relatives we would like to include in our study. Developing criteria for identifying kin to be followed up proved to be the selection task that we found most difficult and we will discuss it in detail in this section.

Extract B shows how our initial thinking on this issue developed.

Extract B: Notes from Record of Planning Meeting 23/9/86 (recorded by Jennifer Mason)

Interviews with Members of the Kin Group–Discussion of Rationale and Strategy

Whose Kin Group Do We Want to Study?

We discussed the pros and cons of either using our respondent as the central person and examining their kin group, or of using them as a way into a variety of kin groups. We decided on the former strategy because:

we will then have survey data on all the people whose kin groups we are examining, because they will all have been survey respondents.

there is a logic of selection whereas if we were to get deflected into other kin groups it would all become very haphazard.

interviews with relatives will thus be used to elucidate our survey respondents kin group and the negotiations within it.

Which Relatives to Select

After a fairly detailed discussion we decided to use the following selection principles:

we will try to interview relatives with whom negotiations have taken place (i.e., where our survey respondents has told us about these).

we will exclude those in "crisis situations" on ethical grounds.

people with whom there has been close contact at some time in the past, i.e., where there has been some kind of negotiation, but where there is now little or no contact.

This is a way of selecting people with whom we might expect our survey respondent to be negotiating, but where negotiation appears to be absent, *without* necessarily limiting ourselves to primary kin.

By following these selection principles we will be being more precise about who we follow up than if we simply interviewed all "significant others" (i.e., significant from our survey respondent's point of view).

How Many Interviews With Each?

As with all the issues we discussed today, we had implicit assumptions and rationales about how many interviews we would want with different types of respondent, and once we had made them explicit we came up with the following:

With our survey respondent we want to know about her/his relationships in all directions, because it is her/his kin group we are studying. With other relatives, although we will have to understand something about their relationships in all directions in order to understand their relationship with our survey respondent and his/her kin, we are interested chiefly in their relationship with her/him. Therefore, *generally we will want:*

one interview only with relatives of our survey respondent, *except:*

where relatives are clearly "significant others" as far as she/he is concerned, in which case they are likely to have more to say about relationships with her/him. In these cases we will want two interviews.

where relatives fall into either our divorced/remarried or young adult categories, in which case we will want two interviews to further our understanding of issues relevant to these categories.

As far as interviews with spouses are concerned, we will only want one because we are less interested in the conjugal tie than in other aspects of kinship. Furthermore, it is neat and tidy to finish the third interview with our survey respondent at the same time as finishing the one with the spouse (i.e., where these interviews are conducted simultaneously). Also, the interpersonal point about feeling awkward because we (including Social and Community Planning Research), will have made three visits to the household by this time.

At this stage we had no idea how successful we would be at gaining access to relatives: we thought it quite possible that our interviewees would decline to pass us on, and even if they agreed, that relatives might refuse to be interviewed. Beyond that unknown element, our major concern at this stage was to decide upon the appropriate balance between getting very detailed information from a very small number of kin groups, and including a range of different experiences in our study. Given the practical limitation of having budgeted over 120 interviews, we could not do both.

Looking back on our research diary from that period, we seem to have resolved that issue (although this is not spelled out in the notes) by going back to look at the purpose of studying kin groups, that is, by working through from the logic of our theoretical ideas and the research issues upon which we were focusing. Certainly our notes do record that we clarified that our purpose in studying kin groups was to understand the process involved in negotiations between kin over financial and material support. We, therefore, felt that we needed to include a range of experiences to help us to generalize about these processes. On the other hand, the whole purpose of studying kin groups rather than individuals was to get accounts of the same issues from different parties involved, and we needed to try to interview enough people from each kin group studied to give us a rounded picture.

We decided to try to follow up the kin group of those people where the initial interview had revealed examples of negotiations between relatives over issues concerning financial or material support. In that sense we were selecting for more intensive study the kin groups of people who "looked interesting" in relation to the issues that we wanted to study. While in the end we stuck to that strategy, it was this particular issue that subsequently gave us cause for concern as we shall explain shortly.

The more practical issues to be resolved at this stage concerned how to define a kin group and how many interviews to have with each person. Our reasoning on both these issues can be seen in Extract B. We decided that in each case we would treat the person who had been in the survey population as the center of a personal kin network, and would focus on that person's significant or close kin as she or he defined them, rather than trying to be passed ever outward along chains of kin. Having taken that decision to focus on clearly defined kin "groups" centered on an individual who had been in the survey, the question about how many interviews to conduct followed fairly straightforwardly: we would do two more with the key person and one each with his or her relatives since our aim was to generate a detailed set of data on the group which centered on our key person. We anticipated that we might sometimes want two interviews with certain relatives for special reasons, but in the event we did not do this.

The disadvantage of this strategy is that it closed off the possibility of following through other relationships that looked relevant to our concerns if they did not involve the key person in the kin group. This did indeed occur. There were several instances where we interviewed a sister or a cousin of our key person who would themselves have made excellent subjects for detailed study. Extract E (see below) shows that we did go on noticing these examples and paused to consider whether our strategy should be changed. However, we continued to reason that we were prepared to sacrifice those possibilities in order to create a systematic selection strategy that would enable us to study a range of different situations. Other researchers might have taken a different decision but we felt any other approach would have led us into a series of ad hoc decisions which, in the end, would be difficult to justify.

PROBLEM 1: COMPARABILITY

Equipped with a reasonably coherent selection strategy, we, therefore, began identifying interviewees whose relatives we would like to follow up. However, we felt a continuing concern to formalize and crystalize our thinking on these issues. The first issue that concerned us was comparability. In each case we were making decisions for good reasons, and following the principles that we had articulated, but we were concerned that these might result in a set of interviews that could be difficult to handle as comparable cases. We would have two interviews with some people, and one with others; with some people we would be following through their relatives and with others we would not.

We resolved this by deciding to turn an apparent weakness to good effect. Extract C shows how we reasoned this through.

Extract C: Janet Finch's Research Notes for 27/10/86

The principle of *instances rather than cases* as the basis of comparability seems to be particularly suited to this project, since we are searching for a range of experiences and social phenomena, and we don't know in advance of the first interview which are present for a given interviewee. We concentrate on those which we do find, but that means that the idea of comparability between cases does not fit this study in any event. Because people have different experiences, we have data on different topics from each person. The principle of taking "instances" as the basis of comparability in a way just goes one step beyond saying that we will have to count up how many interviewees have had a given experience, and that this will be less than the total number in the study. To put together instances of an event from each "side" of it, plus accounts given by third parties gives a more rounded picture. We will need to be able to justify third party examples in terms of taking these accounts *as seriously as* accounts of personal experience, but of course *not the same as* them. Third party accounts are important to us precisely because people are distanced somewhat from the circumstances and therefore they reflect a more "public" view of the situations, reflecting something of how the public morality of obligations gets applied in particular instances, I suspect.

This approach is of course *not the only way* in which we should use examples from our data. I would anticipate that we will want to present it in different forms, and that another obvious one is extended discussion of individual cases (which might be individual people, or kin groups). Obviously the people on whom we have got the most data are likely to be the most suitable candidates for this, although that might not always be the case. My idea about comparability of instances is not to close off other methods of analysis, but simply to see a way through the problems created by the need to present some of the data in aggregated form (i.e., not *just* as a series of individual cases) and the particular issues of comparability that we seem to be building in.

Our idea was that much of our analysis should proceed, not on the basis of comparing each individual person or "case" with another but on the basis of comparing "instances" or examples in our data of particular circumstances in which we are interested. To take one example: people who have temporarily moved back into a parent's home after divorce. Instead of simply counting the number of people interviewed who themselves have moved back and comparing their experiences, we could search in our data for all examples of this happening: people who themselves had done it, people who had been the "receiving" parent, examples given of third parties doing it. Of course, for some of the instances we would have much more detailed data than for others, and for some we would have accounted from more than one party to the arrangement. Some would be personal accounts and some would be second hand accounts. That variability obviously would have to be acknowledged in the way we used the data, but in principle we could draw together a range of instances in this way from wherever they occurred in our data.

The principle of instances rather than cases as the basis of comparability seemed to be particularly suited to our project, since we were searching for a range of experiences and social phenomena, and we did not know in advance of the first interview which would be present for a given interviewee. Again, this comes back to analytic rather than enumerative induction, because it is based on the validity of instances or processes, rather than the representativeness of the sample, as a means of generalizing and of making sociological statements. Furthermore, it illustrates ways in which we were linking our strategies at this stage to ideas about how we would use and write up the data in our formal analysis.

PROBLEM 2: GOOD INFORMANTS

We resolved the question of comparability to our own satisfaction but fairly soon after this we began to be concerned about another issue connected with selection criteria for following up kin. The core of the problem here was that we were worried that we were being seduced into following up the kin groups of people whom we found interesting to interview and who made it easy for us to spot situations apparently concerning negotiations in their kin group. Some of our respondents made very "good informants" in the sense that they talked about their families using concepts that were quite close to our own, whereas others presented material in a more bland way that did not highlight issues like reciprocity, conflict, compromise, working things through, talking things out, and so on. However, this latter group might well be involved in negotiations—in the broad sense—about kin support. Were we in danger of missing the full range of experiences open to us by tending to follow up the people who were—on our distinctive definitions—most articulate and interesting?

Although the detail will vary in different projects, this must be an issue commonly faced by field researchers who want both to use "good informants" and also to produce a rounded picture of the situations they are studying that does not systematically exclude certain kinds of informants or experience.

Extracts D and E show our attempts to resolve this issue. We discussed it on several occasions without getting very far before we decided that Jennifer Mason (who had originally spotted the significance of this issue) should spell out the nature of the problem on paper (Extract D). Janet Finch then responded (Extract E). As a procedure we found this a helpful way of moving beyond our verbal discussions where, on this issue in particular, we had been tending to go round in circles. In these notes, we had reached the stage where we were able to discuss concrete examples of interviews that had been completed. In the extracts these are referred to by their interview numbers.

Extract D: Jennifer Mason's Research Notes 3/11/86

Further Thoughts on Following Up Kin Groups

I have been concerned lately about the implicit criteria I am using during and after interviews to make decisions about whether people's kin groups are worth following up. After talking with Janet we agreed that I should try to make my worries explicit by writing them down.

My main worry has been that there is a danger of only following up the kin groups of articulate respondents—"good informants"—for example 01 or 110. I think my problem is that respondents like this approach some of the issues we are interested in in an analytical way—they are able to reflect and philosophise about their kin relationships—which means that they draw out interesting situations, relationships, etc. to tell us about. In fact, in both of these cases, there are situations which we could presumably identify as interesting even if they were not articulated to us in this way: e.g., sharing accommodation. But I suppose I am concerned that conceptual issues might sway us into following up a kin group—e.g., things like reciprocity, independence and dependence, conflict and tension, giving and lending—and that not all of our initial contact respondents will articulate these. However, their behaviour and kin relationships may still be bounded by/governed by these sorts of issues. So, for example, 101 did not make sharing accommodation sound half as interesting as 110, or 106.

These worries were thrown into relief a bit for me in considering 108 because I had just about made the decision during the interview (exactly as I had made the converse decision with 110), that her kin group would not be worth following up. Yet "objectively" there seemed to be some interesting features: e.g., her relationship with her mother, the effective dissolution of the family home when our respondent bought a house with her boyfriend and her mother moved into warden assisted housing at the age of sixty, examples of cohabitation vs. marriage, her mother's role in caring for her parents before their death a couple of years ago, her mother's widowhood at the age of forty and her consequent retraining and employment as a book keeper, our respondent's consequent close relationship with her grandparents, her determination to marry on the same day as her grandparents, using her grandmother's wedding ring, etc! But all of these things were not articulated in a way which made them sound inherently fascinating. Would 01 have made them sound interesting for us?

I think my nagging doubt is that if we are not careful we will systematically exclude certain types of kin groups, or at least the kin groups of certain types of initial contact respondent, i.e., those who do not reflect or are not fascinated by the intricacies and ambiguities, etc. of family relationships and/or those who are unable or unwilling to articulate these.

Extract E: Janet Finch's Research Notes 5/11/86

Following Up Kin Groups: My Response to Jennifer's "Further Thoughts"

We are trying to use 108 as the focus for deciding more clearly our principles for who not to follow up. But when I listened to this tape, I found it rather interesting, and can identify a number of issues which might well make it worth following through her kin group. (Issues listed in detail at this point.)

So for me, 108 doesn't perhaps present quite the perfect example of the dilemma which Jennifer wrote about, because I think I find it intrinsically more interesting. It may be that there is a general lesson to be learned here: she did the interview, but I have only listened to the tape. It may well be that it is easier to listen for and find interesting issues to pursue when you can listen to the tape without having been influenced by the nature of the interaction. If the interview interaction was difficult, or even just not specially exciting, it may well be more difficult to get enthused about it afterwards than it is for someone who comes to it fresh. So that probably means that we should certainly involve each other actively in any decision *not* to follow up a kin group—whichever of us has not done the interview may be able to spot more interesting possibilities.

My conclusion therefore is that we should probably follow up the kin of 108 working on the criteria which we have already established. But Jennifer's note about the danger of not following up less articulate respondents does convince me that there is a potential problem which I had perhaps been a bit slow to recognise.

What worries me now is: if we recognise that some people's kin groups may be more interesting (to us) than they seem at first sight, are we *ever* going to have a reason for not following up (other than the separate sets of reasons to do with not probing around in crisis situations, etc.)?, i.e., does this effectively amount to a decision to follow up everyone whom we reasonably can? In some ways that would be the easiest strategy to operate and to justify, but I don't feel wholly comfortable about it. I think I am worrying mainly about the best use of our limited resources, in that the more kin we follow up, the fewer kin groups overall we can study. Since material from kin groups (where we do succeed in getting it) is essentially bound to be used as case study material, because we are not attempting to select groups in a way which would make them comparable with each other, then I suppose I have a niggling feeling that I want to get the *best cases* which we can, and which will enable us to understand social interaction and social process but which of course will not in any sense be representative. I don't feel inclined to shift from the strategy of treating our initial contact as the "centre" of a kin group and not to keep snowballing on infinitely with her relatives, and then theirs. To have a cut off of that kind is bound to produce some examples of individuals whom we would have been very happy to see as the contact person themselves (a recent example for me would be 08, who is the cousin of 01, who has a very interesting situation in her own right) but I think I accept that as a consequence of our strategy.

The main way in which we worked through to a solution of the problem of good informants was to capitalize upon having two researchers working collaboratively. Since only one of us was normally present at an interview (and if we were both present, only one of us took an active role and the other was purely an observer) we found that the person who had not conducted the interview was often able to see more clearly the merits of following up a particular case. That person was less likely than the person who had done the interview to be over enthused by a particularly good interviewee, or to dismiss the situation of a more difficult interviewee as being not worth following through. In other words, the other's judgment was unlikely to be clouded by issues connected with having had the responsibility for maintaining the interview as a successful *social* interaction. In this way we were able to take decisions on a case by case basis and to continue to select certain kin groups for detailed study, but not others. It can be seen from Extract E that at this point there was some danger that we should slide back into following through all possible cases, on the grounds that it was just too difficult to distinguish between one and another. But the process of working through the issue with each other and on paper enabled us to confirm that our selection strategy remained appropriate.

From this point onward we adopted a procedure where we always documented the pros and cons of following up the relatives of each interviewee individually. After each interview, but prior to transcription, the person who had conducted it listened to the tape, and produced a family tree, summaries of information given about relatives, and a life story chart. This material, together with the tape, was then passed to the other and formed the basis of a joint discussion about following up kin. This means that in each case both of us listened to the interview tape, and took the next decision collaboratively on the basis of a preliminary analysis. In this way, we felt that we could ensure that we were being as systematic as possible in our choices.

STOCK TAKING EXERCISE

Alongside this developed a more cumulative strategy of discussing each case not just on its own merits, but as part of a growing data set. In this way we tried to keep an eye on the range of experiences that we were studying, and to identify obvious gaps. We formalized this, about halfway through the interview stage, in a stock taking exercise. Again, this was premised upon analytic induction, the logic now being that we should both plug the gaps and begin to seek for "negative instances." As Hughes has pointed out, analytic induction involves:

> A strategy which calls for the investigator to search deliberately for instances that negate his (sic) hypothesis and, using these, to refine the hypothesis further. . . . In practice, the process of analytic induction proceeds by formulating a rather vague generalisation and then revising it in the light of contrary evidence, so that there is a continual process of redefinition, hypothesis testing, and a search for negative cases until a point is reached where a universal relationship can with some confidence be established (Hughes 1976, p. 128).

Our stock taking involved a preliminary but systematic categorization both of characteristics of people and kin groups in the study, and instances of kin support and negotiation in the interviews done so far (41 completed and 17 firmly arranged, out of a target number of 120). On the basis of this, we were able to assess and modify our strategy for the second half of the qualitative stage. Janet Finch did the detailed work of itemizing and categorizing the range of situations already present in our data, and Extract F comes from the record of the meeting where we discussed this stock taking document and used it as the basis for the last major stage in refining our selection strategy.

Extract F: Notes From The Record Of Planning Meeting 14 and 15 January 1987 (recorded by Jennifer Mason)

We used Janet's "taking stock exercise" as a discussion document for our meeting, taking each point in turn. The following are the major points arising:

1. *Categories to Include*

   We decided that, overall, our present strategy for sampling is working fairly well, but that we need to refine it a bit to ensure that certain categories of people are included:

   a. Men. We agreed, perhaps a bit reluctantly (!) that we cannot simply continue to rely on being passed on to men via the kin groups of our "key" women. We need to select some men who have also been survey respondents, not least so that we can then gain access to their kin groups, hence not filtering out this possibility at the start by only selecting women. We acknowledged that this would raise again the unresolved problems of personal safety we discussed a few months ago. We decided that the best strategy would be to go together to interview men and, where that was not possible, to approach a male colleague with experience of field research with a view to his being a "minder." We talked about the possibility of employing a man to do the interviews with men, but agreed that this would be an inadequate substitute for us, given our familiarity with the objectives, perspectives, data already collected on the project.
   b. We agreed that unemployment was an important enough contemporary issue, with implications for our work, for us to include some unemployed people in our sample. Although we cannot tell who is currently unemployed from the questionnaires, which are now a year old, we agreed that we could select people who were unemployed at the time of the survey, and had been for some time, e.g., over a year. If it transpired that they were no longer unemployed such people would nevertheless have experience of a fairly lengthy, and recent, period of unemployment. We agreed to confine this to the under forties or fifties, to prevent the conceptual difficulties with unemployment in later life.
   c. Ethnic Minorities. We agonised over this, feeling that in some ways it would be racist to exclude them, but also to include them on different terms. We agreed that if we were to include people from ethnic minorities they should certainly be people who fitted into our two main sampling categories: young adults and divorced/remarried. Finally we decided that we would get a list of the people involved, and literally take out the questionnaires and examine them to see just exactly what we have got, and what the nature of the situation is. We also agreed that Janet would approach a personal contact with a view to our interviewing him and his wife as a sort of pilot interview. This would enable us to see if our questions made sense, and to discuss with them what sorts of modifications would be appropriate for the different ethnic minorities.
   d. Social Class I, IV and V. We reaffirmed that we want to get a range of experience in our study, whilst also not making any claims as to the representativeness of our

qualitative study group for the general population. Thus, we agreed there was a need for us to gain more people from classes IV and V, to offset the clustering we have at the moment in the middle (II and III). Weight is added to this when we looked at the housing tenure distribution, and our overwhelming bias towards owner occupation. We felt that a conscious attempt to gain people from classes IV and V would help offset this. We also agreed that our survey data made Social Class I (men especially) look interesting enough to warrant a conscious selection strategy here too. Weight was added to this by Janet's "hypothesis" that the continuance of "friendly and civilised" contact following divorce might be a middle class phenomenon (nicknamed the Posy Simmonds phenomenon).

e. Divorce and Widowhood. Janet's suggestion that widowhood might help to throw some light on our understanding of divorce seemed compelling, and we agreed that we should try to include some widows—especially those under fifty years old where this is less common and in a sense more comparable with divorce (does not conflate issues of ageing and widowhood, etc.). We agreed that divorce continued to be a worthy focus and sampling strategy for us, not least because our divorced survey respondents have led us into kin groups displayed a good spread of other "situations" we are interested in. If Janet's "Posy Simmonds" hunch about middle class divorce is right, then our SCI respondents might prove interesting here too. We talked about the possibility of refining our strategy of centering on our "key" respondent as far as divorce was concerned, so that for example if we were to discover a divorced person in their kin group we could possibly follow them up in their own right, that is by treating them as another key person with a kin group. We decided that this would be perfectly valid, and indeed that we could "look out" for some of our other categories and situations in this way too. Now that we are almost halfway through the qualitative bit, and are in a position to reflect on where we are going in the light of where we have been, we felt that there was less danger of losing our focus than there might have been last year in a strategy which allows us to follow up someone who is not necessarily a key figure in our initial contact's kin group, but who is, for example, divorced and of interest in their own right.

f. Elderly. We agreed that we wanted to keep on the look out for elderly, and particularly fit elderly, people in the kin groups, but not to modify our selection strategy in this respect. One of the problems with doing the latter is the danger of our crashing in on a crisis situation which would, in any event, lessen our potential for following up kin given our strategy of non intervention in crises.

g. Step Children and Step Grand Parenting. We agreed that given our interest in divorce and remarriage, this was actually a fairly central issue and we should give it a higher priority. We discussed ways of identifying step parents from the survey questionnaires—i.e., by choosing people who are (divorced or) remarried and who have children listed under "spouse kin." We agreed that we should look out for step parenting situations in the kin groups, but also at the selection stage in the questionnaires.

2. How Many to Sample For Each Category

We agreed, given the fairly heavy commitment of our resources involved, that it would be acceptable to treat men as a "minority group" in our sampling strategy. Partly, we felt this justifiable because our female respondents *are* yielding men in their kin groups. We agreed that we should continue sampling until we have achieved ten men—five divorced/remarried, and five young adults. These ten men can include people in Social Class IV and V, and unemployed (which in fact they do), and we agreed that we should monitor refusals very carefully during this phase, so that where men in these categories refuse we can replace them with other men in these categories.

That would leave us with the following to achieve: ethnic minorities, Social Class I, adult stepchildren/parents, young widowed. We agreed that we would be best to leave until later the sampling decision about young widowed and step children, when we will be in a position to see what we have achieved in these respects from the kin groups. We decided we should target about three or four SCIs, probably men but they could be women (Janet has one potential in her five male divorced/remarried survey candidates). We agreed that we would be lucky to get three ethnic minority initial contacts, at most.

Most of these precise decisions about sampling numbers cannot be made very effectively at this stage, and we agreed that a good strategy would be to allow ourselves a further "taking stock" exercise when we can assess how well "represented" our categories are, and whether there is a case for including other groups/categories/situations. When we are at a stage where we are beginning to feel confident of ideas/theories being generated, we could therefore adjust the sampling a bit in line with the logic of analytic induction.

At this point we were able to be much more focused about selection issues than was possible at an earlier stage. While we were pleased to be able to confirm that the strategy we had been pursuing was generating the kind of data we had hoped for, we were able to engage in some fine tuning. For example, we came back to the categories of respondents whose experiences we wanted to include, but where we had decided at an earlier stage to delay a decision. Thus, we decided actively to seek out examples of people who had been unemployed since we were not picking up examples of these in our existing strategy; but by contrast we concluded that we did not specifically need to seek out elderly interviewees, since we were successfully including their experiences through following them up as relatives within our existing subgroups. We also finally took a decision about men, confirming that we needed to include some as the focal person of a kin group (not just as relatives of women) and resolved the practical problems of security by opting for the labor intensive strategy of accompanying each other to initial interviews with men about whom we had no information beyond the survey interview. We also agreed upon a strategy for selecting people from ethnic minority groups, leaving open the possibility that we might go beyond our two major subgroupings in the case only, to make sure that some non-white experiences were included in our study. This arose solely from the fact that we had a very small number of survey respondents from whom to make a selection; but in principle we were able to confirm that our basic strategy of selecting from the two subgroups of young adults and divorced/remarried was leading us to examples of negotiations within families, and all other selections were made from *within* those two subgroups.

Another slightly different issue that emerged from our stock-taking exercise was that we had so far been interviewing people from a rather narrow social class range, namely, from the middle of the range as defined in orthodox terms. From the point of view of reflecting a range of social experiences in our data, it seemed important to broaden that and we agreed that we should seek out respondents within our main subgroups who fell into classes I, IV, and V. The value of having undertaken a systematic stock-taking exercise at the midway point in the fieldwork is very clear in this instance, since neither of us had realized that our interviewees were bunched in this way. If we had relied upon our informal and intuitive knowledge built up in the course of interviews, we would not have identified this problem until it was too late to do anything about it.

The principle of analytic induction is very explicitly followed in another set of decisions that we took at this time concerning our interest in families that had been reconstituted through divorce and marriage. In the interviews that we had already completed we had plenty of examples of renegotiation of relationships with the person's own relatives, but very few of continuing relationships with relatives of the former spouse. The common pattern seemed to be to cut off contact completely. The "rather vague generalization" with which we began (to put in in Hughes' terms) was that there would be circumstances under which relationships with former in-laws would continue in a renegotiated form. The data that we had collected in the first half of our fieldwork suggested that we should modify our hypothesis to: active relationships with in-laws continue in a renegotiated form after divorce only in unusual circumstances, if at all.

Thus, our data were helping us to modify our theory and that in turn enabled us to test our revised theory further. We decided to do this in two ways. First, we would search for negative instances, which in the context of our modified hypothesis meant that we would seek out those situations where we were *most* likely to find continuing relationships after divorce. In discussing where to look for these we brought to bear our wider knowledge of social theory and of other studies and decided that the desire to continue "friendly and civilized" relationships after divorce is probably a phenomenon associated with the intellectual middle classes. We christened it the "Posy Simmonds" phenomena in our notes: *Guardian* readers will be familiar with this kind of "civility" in the cartoons of Posy Simmonds. This confirmed that we should specifically select some interviewees from Social Class I.

Second, we decided to try to refine our theory further by testing out whether the process of cutting off from in-laws is a consequence of divorce specifically or whether it is a result of the tie that previously bound the people together having been removed. If it were the latter, people who had been widowed would undergo a similar process to the divorced; if the former, the pattern of relationships with in-laws after divorce or widowhood would be very different. This reasoning led us to a decision to select some "young widows" (of either sex) for interview, to test out this distinction by comparing widows and divorcees whose family circumstances were otherwise quite similar.

Extract F shows that we translated these modifications into a selection strategy for the second part of the fieldwork in which we were able to be quite precise about the numbers we were seeking in each category. In terms of the balance between being systematic and being flexible it is clear that we were able to be systematic in a much more overt sense at this stage than we had been at the beginning of the process. But we were still able to retain a degree of flexibility as our notes indicate, we built in the possibility of further stock taking at a later stage. We did in fact repeat the exercise when we were about three quarters of the way through our interviews—this time Jennifer Mason doing the itemizing and categorizing but we made no significant changes at that point.

CONCLUSION

In telling the story of our own project we have made a number of points about how the principles involved in theoretical sampling can be applied and we will not repeat them here. We make no special claims to methodological virtue but we think that it is quite possible to produce a selection strategy in field research that is systematic rather than ad hoc, while maintaining a level of flexibility that is essential within this research paradigm. We have shown how we selected cases to study for their theoretical significance, worked through problems associated with comparability, applied the principle of analytic induction, and made systematic appraisals of the data that we were generating as we went along—all of these within the normal practical constraints of money, time, and, in our case, a concern about the personal safety of the researchers.

We shall conclude by highlighting some of the more general principles that can be drawn from this description of our research process.

1. It is clear that analysis of some kind is constantly taking place, and forms the basis for decisions about strategies, within the overall parameters set at the beginning through a particular theoretical perspective. Different levels of analysis can be relevant here, for example, listening to interview tapes, making a preliminary assessment of each case, itemizing and categorizing characteristics and situations. Preliminary forms of analyses such as these are the raw materials from which informed decisions are made. Theoretical sampling, therefore, encompasses a good deal more than processes generally considered to constitute sampling.

2. Leading on from the first point, this means that decisions made on this basis are not ad hoc. Rather they are both situated and informed. Some decisions simply cannot be made at the very beginning of the research enterprise without loss of theoretical and data sensitivity, yet each time a decision is made it is important to be clear about the principles underlying it, the reasons for it, possible alternatives and so on. On the one hand, informed decisions are part of a process of sharpening or modifying—underlying principles leading to theory that is grounded in data. On the other hand, this is only possible

because to recognize that informed decisions have to be made continually is to acknowledge that the research process takes the researcher through changing contexts. These result from the data being generated, and from continuing exposure to other researcher's theories and findings in relation to her or his own.

3. The implication of this is that delaying some decisions until a later stage of the research process, rather than taking them all at the beginning, is a positive rather than a negative feature. However, this is not a license for the researcher to be ad hoc, and to make decisions simply off the top of her or his head. In essence, what must be gained from any situated description of informed decisions is a lesson in how to be systematic. This is rather more of a challenge than to be systematic in a positivistic sense, because informed decisions made in-progress can easily appear ad hoc and inconsistent if the researcher cannot be entirely clear about the changing contexts of those decisions, their purposes and consequences, and the principles underlying them. In the absence of this vital contextual information, it is dangerously easy for researchers to telescope decisions made into a positivistic model, by suggesting that they had sorted out most of the issues at the very start.

4. Being systematic in preliminary analyses of one's data in this way represents the beginning of a cumulative development of principles of analysis. Therefore, as well as it being important to record both decisions and contexts for an expose of the practicalities of theoretical sampling, these very records form excellent documents for use in the early stages of the formal analysis of data.

5. Collaborative working methods are, we have found, a positive bonus in all of this. We have been able at each stage to have real discussions about decisions and issues as they occur, and kept records of these discussions as well as our individual endeavors. However, if some of the processes involved in being systematic seem more obvious in a collaborative working context, they do not have to be exclusive to it.

ACKNOWLEDGMENTS

We would like to thank Bob Burgess, Caroline Dryden, John Hockey, and Sue Scott for reading and commenting on an earlier version of this chapter. Also thanks to colleagues present at the meeting of the qualitative methods study group in the Department of Social Administration at the University of Lancaster, where we discussed an early version of this paper: Nick Derricourt, Joy Foster, Anne Williams.

NOTES

1. This study is supported by a grant from the Economic and Social Research Council, 1985-89. The total grant was £121,000 of which about £50,000 represented the cost of the fieldwork and data processing for the large-scale survey which formed part of the study.

2. The survey was conducted in the Greater Manchester area and was based on the electoral register in forty cluster sampled wards. A response rate of 72% was achieved, making a total of 978 completed interviews. At the end of the questionnaire respondents were asked, "It is possible that a researcher on this project might want to come back in some months' time. Would you be willing to give another interview?" Eighty-five percent said they would be willing.

3. The survey fieldwork was organized and conducted through Social and Community Planning Research, and we would like to gratefully acknowledge Gill Courtenay's contribution and support in this stage of the project.

REFERENCES

Baldmus, G.

1972 "The Role of Discoveries in Social Science." In T. Shanin (ed.), *The Rules of The Game*. London: Tavistock.

Bell, C.

1977 "Reflections on the Banbury Re-study." In C. Bell and H. Newby (eds.), *Doing Sociological Research*. London: Allen and Unwin.

Bertaux, D.

1981 *Bibliography and Society*. Beverly Hills, CA: Sage.

Burgess, R.G.

1984 "Autobiographical Accounts and Research Experience." In R.G. Burgess (ed.), *The Research Process in Educational Settings: Ten Case Studies*. Lewes: Falmer Press

Finch, J.

1987 "Family Obligations and the Life Course." In A. Bryman, B. Bytheway, P. Allatt, and T. Keil (eds.), *Perspectives on the Life Cycle*. London: Macmillan.

Glaser, B. and Strauss, A.

1967 *The Discovery of Grounded Theory*. Chicago: Aldine.

Hammersley, M. and Atkinson, P.

1983 *Ethnography: Principles and Practice*. London: Tavistock.

Hughes

1976 *Sociological Analysis: Methods of Discovery*. London: Nelson.

McKee, L. and O'Brian, M.

1983 "Interviewing Men: 'Taking Gender Seriously.'" In E. Gamarnikow et al. (eds.), *The Public and The Private*. London: Heineman.

Platt, J.

1976 *The Realities of Social Research*. Brighton: Chatto and Windus/Sussex University Press.

Porter, M.

1984 "The Modification of Method in Researching Postgraduate Education." In R.G. Burgess (ed.), *The Research Process in Educational Settings: Ten Case Studies*. Lewes: Falmer Press.

Schwarz, H. and Jacobs, J.

1979 *Qualitative Sociology: A Method to the Madness*. New York: Free Press.

Stenhouse, L.

1980 "The Study of Samples and the Study of Cases." *British Educational Research Journal* 1-6.

Znaniecki, F.

1934 *The Method of Sociology*. New York: Farrar and Reinhart.

IT'S NOT A LOVELY PLACE TO VISIT, AND I WOULDN'T WANT TO LIVE THERE

James M. Henslin

INTEREST, PLANS, AND REALITY

When the homeless were first reported in the mass media, I found them only of slight interest. The homeless seemed merely another passing phenomenon, albeit an interesting one. They struck me as depressingly similar to so many other features of modern society, making yet another caustic comment on the injustices of contemporary industrial life. However those accounts of their plight on television and in newspapers might arouse my pity, the homeless remained strange, appearing from where I was sitting, merely remote, impersonal representatives of yet another social and economic aberration.

A short time after the first reports of the homeless appeared in the media, however, my path crossed with that of an individual who was destined to change my comfortable aloofness—intellectual and spatial—and, for good or ill, to

Studies in Qualitative Methodology, Volume 2, pages 51-76.

ISBN: 1-55938-023-3

affect profoundly the direction of my sociological research. For this person, the homeless were no conceptual or media abstraction. Rather than divorced from life, the homeless had become his consuming passion in life.

Several years earlier, as he was studying cold, intellectualized formulations of God at a seminary, this man had become deeply bothered by the plight of people without homes. Because these abstractions about God did not appear equal to the suffering he witnessed, he and his wife opened their own modest home to these strangers. They had nothing to offer but the floor of a mobile home, and that is what they provided. Without institutional support, for his church apparently had better things to do, he and his wife ventured onto the streets in search of the hungry and freezing. After wrestling with the purposes of being trained in theology, he resigned from the seminary and he and his wife made those forays into the streets their full-time work. Now, after a dozen years with the homeless, they were operating overnight shelters and rehabilitative farms for the homeless—as well as the only television station in the world run by and for the homeless.

When this individual, Larry Rice of St. Louis, Missouri, found out that I was a sociologist and that I was writing a textbook on social problems, he asked me to collaborate on a book about the homeless. He felt that my background might provide an organizing framework that would help sort his many experiences and observations into a unified whole. During our attempt at collaboration, he kept insisting that as a sociologist I owed it to myself to gain first-hand experience with the homeless. Although I found that idea somewhat appealing, because of my heavy involvement in writing projects I did not care to pursue the possibility. As he constantly brought up the topic, however, I must admit that he touched a sensitive spot, rubbing in more than a little sociological guilt. After all, I was an instructor of social problems, and I did not *really* know about the homeless.

The fact that I did not *really* know (in the sense of having first-hand experience with) bank robbers, check forgers, corporate criminals, pornographers, rapists, the insane, and so on, and still was able to teach about them, eventually became irrelevant under his relentless prusuit of this point.

With the continued onslaught, I became more open to the idea. (Or perhaps I should say that I eventually wore down.) When he invited me on an expense-paid trip to Washington, DC, and promised that I would see sights hitherto unbeknownst to me—such as homeless people sleeping on the sidewalks in full view of the White House—firing my imagination, he had pierced my armor through. With the allure of such an intriguing juxtaposition of power and powerlessness, of wealth and poverty, how could I resist such an offer?

And that trip proved to be the significant turning point. From the comfortable life that I had built for myself, living in a small Midwestern American town, enjoying a Victorian home in a "good" neighborhood, and teaching at a university which, with its location in the midst of a sprawling

2,600-acre campus, also was safely sheltered from the "evils" of the city, I was thrust into a disjunctive range of experiences. What I saw and heard were quite unlike the remote televised reports, antisepticized sights and sounds that become almost fictionalized in their very process of disassociative transmittal. In contrast, with my own eyes I saw real men and women sleeping on the heating grates of the Federal buildings—within view of the White House, as promised. And with my own ears I heard people dressed in rags and sitting on the sidewalks talking aloud to no one visible but to themselves.

How could I leave such scenes untouched—or unchanged? Having seen those conditions for myself, both my sociological imagination and my human curiosity were piqued. I now had to know the *why* that underlay those dissonant sights. Who are those people? How do they get there? How do they survive? Do they have any chance of getting off the street? What are their feelings, and hopes and dreams? And, what motivates the people who are helping them?

These questions, and those haunting images, remained with me as I escaped to the embryonic security of my own world. No matter how I tried, I could not throw them off. I *had* to find the answers.

To understand those startling events adequately, seemingly overnight burst upon the western world, I knew that I had to grasp the larger picture, and yet avoid a morass of abstractions that would make me lose sight of what the individual homeless were experiencing.

But how to get those answers? To solve that, I had to wrestle with three major problems of data gathering: the "how" (the particular research method), the "where" (the particular research setting or site), and the "who" (the particular sample within the setting).

Concerning the how: with my own research background, there was no question that I would use qualitative methods. If my research interests had been different, however, such as wanting to know the incidence of illness among the homeless, their income levels and patterns of residence during the years preceding their initial homelessness compared with matched random samples of the American population, and so on, I would have needed quantitative methods. The very posing of research questions, however, precludes some research approaches while dictating the choice of others, and determining what the research questions will be depends on the background of the researcher—who tends to ask questions that can be answered according to his or her particular methodological training, interests, and qualifications, tending to avoid questions that violate those and would indicate some contrary approach.

Because I feel strongly that, compared to quantitative approaches, qualitative methods are more closely tied into the realities that people experience and thus enable the researcher to better comprehend how people cope with their problems, how they attempt to maintain a semblance of order in their lives, and how they make sense out of the stream of experiences that they encounter and create, there was no contest. Everything pointed to qualitative methods. And with my research background, this meant participant observation coupled with interviewing.

The "how" is never that simple to answer, of course, and I still had to decide the particulars that I would fit into the broad framework that we call participant observation. The question of the extent of participation remained undecided, as did the matter of whether the research should be overt or covert. These specifics merged with the "where" and the "who" of the research.

As I wrestled with those questions, reports about the homeless began to originate from many parts of the nation. My imagination became fired by news reports about a tenting community in Houston, Texas, bag ladies in New York City, and tramps in Seattle, Washington. Increasingly I realized that I would not be satisfied with learning about the homeless only in some local area. But could one person do a national study? The only national studies I could recall involved teams of researchers heavily financed by corporations or by government agencies. In order to maintain my independence of topic, timing, and methodology, over the years I steadfastly had refused to apply for research grants. Even when some of my research had come to the attention of an agency of the Federal government and a director of that agency had visited me and had asked me to submit a proposal for "almost guaranteed" megabuck financing, I had refused to participate.

As I ruled out what appeared to me as a time-consuming, cumbersome process of grant applications, one with a highly uncertain outcome, the possibilities of conducting this study the way I desired seemed remote. While I was debating alternatives, however, my university announced that it would offer $3,000 competitive summer research awards. I quickly calculated the cost to myself of accepting such an award (about $8,000 in lost summer salary) and decided it was worth it. After I was granted this award, the specific problem was how I could travel around the country on $3,000.

In doing research not everything works out as intended, and often the researcher must adjust his or her research design to match changing conditions. Frankly, my first attempt to do this research met with resounding failure. Wanting my wife and five-year-old son to accompany me, and wishing to combine sight-seeing travels with research, I purchased a used motor home. (Visions of excursions across the country with an occasional foray onto skid row danced through my head!) This vehicle proved highly adapted to the purpose of sight-seeing and family life. But not so for this research. A conflict arose when it became evident that what we considered scenic were such things as snow-capped mountains, verdant forests, meandering streams, and rushing waterfalls—a far cry from the garbage-strewn strteets and other urban blight of the skid rows where the homeless congregate. And where to safely park the motor home once on skid row? And what should my wife and child do while I was locating and interviewing the homeless?[1] Scratch that one. After a couple of awkward visits, it became evident that I had to drop the skid rows of America from the family tour. Yet that was precisely where I had to go. After a trip through the Rockies, we returned home—and, as they say, back to the drawing boards.

Again, the problem was how to match ideal research goals with the exigencies of the practical world. Able to sell the motor home for what I had paid for it, I still had the original grant to finance the research. I heard about a "fly-anywhere-we-fly-as-often-as-you-want-for-21-days" sales gimmick from Eastern Airlines. I found that their offer was legitimate, that for $750 I could pack in as many cities as I could stand—actually more than I could stand as it turned out, but more on that later.

It was the method itself, participant observation, that became the key for making this research affordable. Obviously, the homeless spend very little money, which dovetailed perfectly with my situation and desires. I was able to stay in the shelters at no financial cost. (The shelters, however, exacted a tremendous cost in terms of upsetting my basic orientational complacencies.) In addition to a free bed and a shower, the shelters usually provided morning and evening meals. Although those meals were not always edible, I was able to count on the noon meal being of quality, and that was already included in the price of my airline ticket.

The question of the "where" and the "who" could not be solved in the abstract. They were always at least partially "in process." Concerning the "where": during the first phase of this research I primarily focused on major cities in the Western part of the United States, later adding cities in other areas during subsequent travels. My purpose was to obtain as good a "geographical spread" as I could. Concerning the "who": Vagaries of sampling are an inherent part of participant observation, but in using this method one need not be submissive to chance only. In this instance, I tried for as much variety as possible, making certain that I had samplings by race, sex, age, appearance, time of day, and, where possible, location within a city. The specific who of the sample, however, depends both upon chance (who is present in a given location at a given time) and the researcher's background (in making one's choice within a particular setting, some people simply look more interesting, more approachable, and so forth).

THE RESEARCHER'S FRONT

In his classic work, *The Presentation of Self in Everyday Life* (1959), Erving Goffman focused his penetrating analysis on the background assumptions of everyday social interaction. Goffman's book created a stir among sociologists and others precisely because he took an unabashed look at the taken-for-granted aspects of everyday life (as well, of course, by the artful way in which he accomplished his task). In delineating the expectations of interactors,

Goffman focused much of his analysis on what he termed personal "front." He emphasized that successful interaction requires the manipulation of front in a way that satisfactorily meets the expectations of those involved in the interaction scene. One's front, stressed Goffman, is made up of three essential components: conduct, appearance, and manner. Conduct is what people do; appearance is how they look; while manner is how people do what they do, including the attitudes they express. To delineate people's everyday manipulation of front, that is their attempts to control the impressions that others receive of them, Goffman coined the term "impression management."

Just as Goffman indicated that impression management is an essential component of everyday life, for we all must attempt to manage the impressions that others receive of us if we are to complete the myriad routines in which we are immersed, impression management is no less essential for succeeding in field research. To adequately manage our research tasks similarly requires a skilled manipulaiton of front: After choosing a feasible role, we must look the part that we have chosen (appearance), make certain that our words and actions correspond with this role (conduct), and also exhibit a presentational style (manner) that facilitates this role. To a great degree, the results of fieldwork depend on the effective alignment of these essential parts of front.

The primary question of front on which my research task depended centered on appearance. I had to ask myself if I should look like a homeless person, perhaps even wear a disguise and "become" a homeless person. Covert participant observation would certainly be a viable way to conduct research on the homeless (Coleman 1988), and with its many appealing features—especially those of being able to participate extensively in the subjects' natural settings and to apprehend unfolding interaction without anyone's awareness of the presence of a researcher—I ordinarily would have chosen that approach.

But appearance must be tempered with purpose, and my purpose involved more than to participate in the world of the homeless.

My basic research questions had expanded. I realized that if I were to adequately apprehend the world of the homeless I also had to focus on their essential counterpart—those who make it their job to work with the homeless. This, combined with my desire to obtain a sample across the nation and coupled with the need to move quickly in and out of cities, meant that I could not be limited to making observations. Interviews would also be necessary. While the observing could have been done well undercover, the interviewing could not.

The need, then, was to be able to mix unobtrusively with the homeless, and yet, at my will, to be selectively identified as a professional researcher. I must at least be able to walk up to people working with the homeless, to announce that I was a sociological researcher, and to have them accept my claimed identity. At the same time, however, I dared not stand out from the homeless. If I were to do research by traveling to unfamiliar cities and going to their skid rows, it seemed essential that I blend in with the ordinary participants of those urban ghettos. If not, I risked two dangers, one professional and the other personal. On the professional side, the observations could become suspect as my presence could change the interactions. On the personal side, I could run the risk of marking myself as a tempting target for muggers.[2]

To fit in, I chose an older pair of levis, a short-sleeved pullover shirt, and a pair of faded running shoes with regular (not running) socks. I knew that I was going to be doing a lot of walking, so the running shoes would also be highly functional. But I wore a cheap-looking pair of shoes, not unlike those that many poor people wear: No emboldened letters, for example, proclaimed an exclusive brand or otherwise made them stand out as expensive.

This costume fit the bill well. (I almost hesitate to call this garb a "costume," as it is actually my everyday "around the house" clothing, but that, of course, is a costume).[3] I blended in. No one on the streets looked at me twice, as if to say, "You don't belong here—you don't fit." In short, my presence went unquestioned.

At the same time, however, to be able to focus on the shelter personnel presented the need to look like a researcher, a seemingly impossible dilemma as this requisite apparently demanded opposing appearances. The solution, however, turned out to be relatively simple. It merely required, in Goffman's terms again, a "prop" that would communicate the role of researcher. Yet the prop had to be one that fit into the setting and would not contradict the role of street person.

I happened to have an old briefcase that long ago I had shoved into the back of a closet. Actually, it is similar to the case carried by pilots—only mine was made of obviously cheap vinyl. It was something I had picked up a dozen years before—for what I no longer remember—one of those nuisance items that is too "useful" to throw away but not good enough to use. I also needed something in which to carry my tape recorder, recording tapes, and a change of underwear, socks, and shirt.[4]

This prop, serving well for the purpose of identity-translation, also blended in well on the street, as no one in the ghetto questioned why I was carrying this satchel-type bag. In addition, its cheap corner stitching soon began to unravel, making it look as though I had just snatched it up out of the trash. Perhaps the best illustration of my fitting into the setting, although not necessarily the role I had chosen, occurred shortly after my arrival in Seattle, Washington. As I stood in a doorway of Seattle's skid row, trying to get my bearings before approaching anyone, a young man came up to me and asked me if I had a certain drug. I said that I was not selling anything, and as he turned away to leave another young man approached and asked how much he wanted. They then made their deal.

When I wished for it to do so, this prop would set me apart. When I would announce to shelter personnel that I was a sociologist doing research on the homeless, they immediately would look me over—as the status I had announced set me apart from the faceless thousands who come trekking through the shelters—making this prop suddenly salient. To direct their attention and help them accept the announced identity, I noticed that at times I would raise the case somewhat, occasionally even obtrusively setting it on the check-in counter (while turning the side with separating stitching more toward myself to conceal this otherwise desirable defect).

But I realize that there was more to my metamorphosis from homeless man to sociological researcher than making that announcement and employing that minor prop. After all, I still looked like the other men, and if it had not been for my alleging the researcher role the shelter personnel would have paid no particular attention to me. This is where the correlary components of front surfaced into prominence—my appearance and manner merging for the role transition. As my turn came in the homeless queue, I noticed that I would change my posture: I would stand somewhat more erect in order to make my presence a little "fuller." My voice became a little firmer, carrying with it greater assertiveness and the communication of professional bearing. I would announce myself by saying "Hello. I'm Dr. Henslin from Southern Illinois University in Edwardsville, Illinois". I would then pause for a moment while the individual looked up at me, and when I had his attention I would then add, "I'm doing research on the homeless, and I would like to talk to you".

This communication turned out to be adequate. Much to my surprise, no one ever questioned who I was. No one even asked to see my identification, which I always had ready.

REALITY SHOCK AND THE IMMERSION PROCESS

Immersion into the world of one's subjects can be traumatic for the reseracher if that world conflicts sharply with one's own. Indeed, initiation into that other world can be so jarring to one's sensibilities that, regretting the folly of one's choice, one immediately longs to flee to the security of the familiar. Napoleon Chagnon (1983), for example, was profoundly shocked when he first set eyes on the Yanomamo: It simply was not part of his ordinary world to be around people who had green mucus stringing down from their nostrils.

Chagnon's desire for instantaneous flight, however, was mitigated by fieldwork realities that made flight impossible—the boat that had deposited him on those strange shores had already departed leaving him stranded with little alternative but to forge on to try to make the best of what he found to be a fearsome, bewildering world.

In my case the transition was considerably more gradual, for which I am thankful. I had arranged to begin interviewing the homeless following the 1984 meetings of the American Sociological Association in San Antonio, Texas.

There I was lodged in a middle-class hotel, one with a lounge, room service, and so on. From the amenities of that hotel and the ambience of the convention, I was abruptly transported to the world of the homeless, albeit with some reluctance.

The first step in the immersion process was to find the homeless. It turned out that there was very little physical distance separating us. Some homeless inhabited the area that I had been traversing from my hotel to the convention site, only they had been invisible to me. One of my first lessons, indeed, was that not all homeless look like winos or bag ladies, that many distinctions between the homeless and "us" are not great.

Not knowing where the homeless were, however, I simply left the hotel and began walking, keeping my eye out for homeless people—and feeling foolish at the same time. Then one of those serendipitous aspects of field research occurred. The first person I stopped to ask if he could direct me to a shelter for the homeless pulled out his billfold and unfolded a list of places that provided free food. Seeing that he had this list, I asked him if he were homeless. I noticed that he became somewhat defensive, but after looking at me for a moment, he replied, "Yeah." I would not have been able to spot that man as homeless as he looked like anybody else on the streets.

I asked him if he were hungry, adding that I was looking for a place for lunch and offered to buy him something. He told me that he had just eaten at one of the places on his list, but said he would accompany me to a restaurant and have a cup of coffee.

My first interview, then, was conducted in the safe surroundings of a restaurant, where I discovered that my subject was a personable individual who related to me a series of hard-luck events. He said that he had just come from public legal services in the attempt to resolve one of his many problems. When I said that I had just begun my research and did not yet know what to believe, he showed me the documentation that verified his story.

After this interview I went to the nearest shelter, to which this individual had directed me. That also proved fortunate for my immersion into a strange world. Indeed, of all the places I visited, this first shelter turned out to be the most impressive of all. It was small, taking in only about 30 men a night. And it was run by caring people who went about their work with a "personal touch," something rather rare in the world of the homeless. Out of deep conviction that their faith in Christ required personal service to the poor, a group of independent, fundamentalist Christians had purchased a downtown home and opened it to the homeless.

I did not know it at the time, of course, but theirs was the best treatment that I would encounter during this research. Even the meals had a "homey" touch about them. Impressed by my credentials and perhaps somewhat concerned about my lack of experience with the homeless, the director suggested that I spend my first night in his office. He then gave me a key to the office, bedding for the couch, a supply of clean towels, and access to a private shower. Able to leave my newly found refuge to conduct interviews whenever I desired, and even to drink coffee at my leisure, I thought that my anxious anticipations had exaggerated the problems I would face.

I always shall remain grateful for that gentle immersion—a sort of half-way house for the stumbling researcher's first night in the field. The next night, however, brought with it culture shock. In Houston, Texas, I visited one of the largest shelters in the country, where in a building that looked as though it once had been a warehouse for objects they now sleep 400 to 500 homeless men a night. In the room to which I was assigned, 55 bunk beds crowded together 110 strangers.

It was not the shelter's larger size and greater impersonality, however, that brought culture shock. It was, rather, its radically different approach to the homeless. For example, at check-in each man was assigned a number. At the exact designated time the man located a bed marked with that number, one that held at its foot a similarly-numbered basket. Each man then undressed at his bedside and waited in the nude until his number was called. Still nude, he then had to parade in front of the other hundred and nine men, carrying his clothing in the numbered basket to a check-in center operated by clothed personnel. From there he walked to a group shower, where, at the time I showered, a young male, with eyes glazed and seemingly lost in reverie, was slowly stroking his distended penis. After showering, but still standing in the nude and surrounded by nude strangers, each man was required to shave, using the common razors laid out by the sinks. Finally, still nude, he took the long walk back to his assigned bed.

This routine burst upon me as a startling experience, one similar to the initiation rituals Goffman (1961) described as characterizing total institutions—perhaps it more accurately can be designated a degradation ceremony, designed to remove one's claim to a proper place in the human landscape (Garfinkel 1956). For those who have gone through basic training in the armed services, such an experience might only trigger memories—that perhaps by now are fond ones at that. For me, however, to parade nude in front of strangers, to be present while a man masturbated, and to witness man after man parading nude, was humiliating and degrading, a frontal assault upon my sensibilities.

Nor was that night spent peacefully. Gone now was my cuddly sleeping partner of the past dozen years. Gone were my familiar surroundings. And, especially, gone was the lock that protected me from the unknown!

Instead, I found myself locked in with them, the unknown!

And who were those people? They were the vagrants, who might be all right but certainly were not my regular associates. They were the middle class, down on their luck. They were the unemployed poor. They were the alcoholics and the drug addicts kicked out of polite society. With none of these would I ordinarily walk in the nude, shower with, or spend the night. Although I had reservations about these subgroups of homeless, it was not really they who bothered me.

The group about whom I had the greatest trepidation was the criminals. Some unknown proportion of the homeless are wanted by the law for a wide range of criminal offenses. Of those in flight I did not really mind the petty thieves or those who may have skipped out on child support payments. What really bothered me were the street criminals—the violence-prone—the rapists, the muggers, and the murderers. "Perhaps none of these men is like that", I tried to tell myself as I "casually" glanced around the room. But my observations failed to provide reassurance—so many of them matched my stereotypes of such persons.

Then my mind insisted on playing back statements made by one of the directors of the shelter. Earlier that day, as I was interviewing him in his office, he mentioned homosexual rapes that had occurred in the dormitories. Then during the interview two men had to be removed from the dining hall after they drew a knife and a pistol on one another. When I told him that I was planning to spend the night and asked him if it was safe, instead of the reassurance I was hoping for, he told me about a man who had pulled a knife on him and added, "Nothing is really safe. You really have to be ready to die in this life."

That certainly was not the most restful night I have ever spent, but by morning I was sleeping fairly soundly. I knew that was so because in the early hours, at 5:35 to be exact, the numerous overhead lights suddenly beat onto my upturned face while simultaneously over the loudspeaker a shrieking voice trumpeted, "Everybody up! Everybody up! Let's get moving!"

Once again, the nude routine. You wait until your number is called, then you parade in the nude to the counter, walking as nonchalantly as possible through the maze of bleary-eyed men. There you retrieve your numbered basket and belongings, file past the men once again on your way back to your numbered bed, where, still on public display, you dress for the day.

This particular sequencing of events, of course, from protective privacy to group nudity, is simply one of the vagaries of field work. My experience with group nudity and my upsetting loss of privacy could just as well have occurred during my first night in the field. But it did not, and by the time it did occur I had already talked to some of the homeless, gaining a sense of them as suffering individuals. My first night's pleasant experience, then, mitigated, at least somewhat, the following night's more disquieting, disjunctive immersion into the routines of the homeless.

DEVELOPING STRATEGIES FOR LOCATING SUBJECTS

In some types of social research, to locate subjects is no problem: They are readily visible and nearby. For example, many researchers find this to be the case with college students who, consequently, tend to be overrepresented in our studies. Others group themselves together in ways that make them highly accessible to researchers, such as by frequenting a certain bar, store, or hotel, or even by living in a particular neighborhood. Some, however, are much more difficult to locate because their connections to one another are tenuous, such as those who are involved in similar activities but who seldom associate with one another (e.g., McCall 1980). Being easy to locate does not say anything about the ease or difficulty of penetrating that world, of course, and high visibility itself can be a defense against intruders (e.g., Thompson 1988).

The homeless are not grouped in formal associations, such as are motorcycle gangs, nor under institutional aegis as are college students. Some of them, however, do frequent their own equivalent of restaurants and hotels, giving us numerous studies focusing on city shelters (cf. Arce, Tadlock, Vergare, and Shapiro 1983; Cimons 1976; Crystal, Goldstein, and Levitt 1982; Strasser 1976; *On The Streets,* 1984).

If the goal of this research had been to focus on shelter providers, even though there is no central registry, it would have been fairly easy to put together a listing of such providers. For each city one would need to locate only a single shelter, as the workers of each shelter know or have a list of other shelters in their city. Usually the Salvation Army will do for starters, and its location is given in the telephone book. In addition, studies published by government agencies sometimes contain a list of shelters (e.g., Bishop 1983).

The goal of this research, however, to talk directly with the homeless in cities around the country, made the task of locating subjects somewhat more difficult, especially as I wished to include homeless people who do not use city shelters. If a map of our skid rows had been available, of course, it would have been most helpful.[5] As it turned out, however, it was the similarities of the developmental stage and physical layout of America's cities that provided the basic mapping strategy for locating subjects.

American cities are in roughly the same stage of development: Having undergone severe deterioration, the downtown area of each city currently is experiencing what many call its renaissance (Clark 1981). Yet each of our large cities still contains a rundown area in which almost inevitably the Greyhound/Trailways bus station is located and where various types of deviants gather, especially transients and vagrants.[6] In spite of its "seediness", this downtown sector-in-transition also boasts a new Marriot, Hilton, or other high-rise luxury hotel catering to the upper-middle class.

I was able to utilize these characteristics of our cities' structure to develop a basic strategy for locating the homeless in cities in which I was a total stranger. After disembarking from the plane, I would immediately head to the airport limousine booth and request the downtown hotel that was closest to the Greyhound. This technique never failed. Inevitably, within two or three blocks of this glass-enclosed, imposing hotel were clustered the homeless—reluctantly tolerated in an area slated for renewal and additional high-rise structures.

EMOTIONAL AND SELF-IDENTITY RISKS

Primarily two threats to emotional well-being and self-identity due to fieldwork are identified in the literature. The first is the risk of field-workers becoming overly intimate with their subjects, thus jeopardizing their objectivity (Pollner and Emerson 1983). The second is the risk of "going native." In this process of radical change, or self-transformation, the researcher may so identify with his or her subjects that he or she remains with them, no longer able to reassume earlier identities in his or her previous world.

Fieldwork also poses a third risk, that of "disjarment," which can similarly threaten the identity with which the researcher began field research. It is not unusual for field researchers to find that their research experiences challenge at least some of their ordinary assumptions of reality. Their intimate experiencing of contrary frameworks of thought and action can hardly leave them untouched. The world one is researching may so differ from one's own, however, and one's immersion in it be so deep, that one's ordinary assumptions of reality are not merely challenged, but those assumptions come to be seen as ill-fitting with life itself. In this experience, which I am calling disjarment, the researcher may come to see his or her earlier assumptions of normalcy as inappropriate to retain. Unlike going native, however, one returns to one's everyday world. But once back, no longer does one feel that the "parts" of that world "fit". And with the disappearance of this "fit" goes a good part of the researcher's own identity.

A fourth threat posed by fieldwork is "engulfment". By this term I refer to another form of being overwhelmed by the research experience. As I flew from one unfamiliar city to another, for me, the skid rows of America were only 24 hours apart. A skid row might be halfway across the nation, but I would be there the next day—with another skid row immediately following the day after, and another the day after that. There I sought out the homeless wherever I could find them. I talked to the homeless in shelters, parks, bars, and restaurants, on streets, buses, and beaches. I did research from the time I got up until the time I went to bed—and I immediately would begin interviewing anew the next morning. When I was not interviewing the homeless, I was dictating my obervations and initial analyses about them.

This constant process of interviewing and recording observations and analyses left me exhausted.

This shuttling between two sharply contrasting worlds also exacted an emotional toll. Flying from one city to another on almost a daily basis forced me for several hours at a time to interpenetrate the world of airport personnel, affluent businessmen, and other air travelers. Not surprisingly, the front that worked so well in the ghetto did not match this more privileged world.

After walking the streets with the rejected and dejected, after sleeping on dirty mattresses among the alienated and despairing, and after taking group showers with masses of strangers and eating meals whose benefits for one's health were often doubutful, I would be transported abruptly into the middle-class world with its sense of orderliness, cleanliness, privacy, and other background assumptions of normalcy. I had no opportunity to carry a suit or other "dress up" clothing, but there I would be sitting among ultra-clean and pressed and processed businessmen wearing their three-piece suits and carrying their look-alike designer luggage and briefcases—matched by female counterparts who closely resembled them.

And there I sat in their midst, carrying my raggedy, unraveling "briefcase" and one piece of shoulder luggage, wearing my scuffed running shoes, faded levis, and scruffy shirt. In addition, with the press of research and the "facilities" to which I was exposed, I would not always smell the best.

This radical disjuncture created a problem in my psycho-emotional adjustment—only one of the many I endured during this research. In defensive reaction, I would force myself to shrug off the various communications that I was "out of place," even attempting to smile within, taking fleeting refuge in the "superior" knowledge that I knew, despite appearances to the contrary, that I "really" fit into this middle-class world. I sometimes also felt a strong urge to reveal my "real" identity to seat mates during flight, another form of reclaiming my sense of belonging.[7]

It was not only the intensity of the research experience and the tearing of identities that bothered me. It was also the type of people with whom I found myself immersed. It is not for nothing that skid rows are infamous for attracting deviants and for being less than desirable places to visit or to live. During the day most deviance was kept surreptitious as the area was shared with office workers and the managerial class. As night fell, however, I would observe the metamorphosis of the city, transforming itself from its more sedate daytime activities into a panoply of deviance. As drug dealers sold their wares the pimps and prostitutes openly awaited their symbiotic counterparts, and muggers their victims. I found myself enmeshed with people who, unashamedly, conversed aloud to themselves, while others, mounted on roller skates, danced to the blaring music of ghetto blasters. Others proudly displayed heads shaved in unexpected places, with the remaining hair, protruding directly outward, dressed in bright, contrasting hues of oranges and purples and reds and greens and blues and yellows. Men in their 40s and 50s, with huge beer guts protruding through open denim vests had skinny, 16-year-old, blond females posed on the backs of their motorcycles.

This constant entering of one unfamiliar urban area after another to solitarily search for homeless people in high crime rate surroundings inhabited by people whose worlds were so dissimilar to mine created a degree of anxiety. One night in San Francisco remains especially vivid in my memory. About 11 o'clock I was in the midst of a scene such as I have just described. Wrapping up the day with an interview with a homeless young man, I sat on the sidewalk with my back propped against a building, feeling much more secure facing outward. After muttering things like, "What we're dealing with is actual treason," "It's about a total witchcraft takeover," "They'll probably kill me," and "You're not going to get away," he stopped speaking and began to stare intently at me. Feeling my body stiffening, I waited, for what I did not know. Without moving his intense stare, he broke the uncomfortable silence by saying, "I know why you're here." He then paused, saying nothing more, yet not taking his eyes from mine. As I looked back questioningly, he abruptly asserted, "You're here to help me, aren't you?" Remaining calm and unmoving, in a slow, measured voice I replied that I was only here to do research on the homeless. Shaking his head, he threw off my answer saying, "No. I can tell by your hand movements that you've come to see me."

Somewhat bewildered and a bit incredulous, I felt highly vulnerable. He then said, "You have to admit that you know my problem, and you came out to help me specifically." As I slowly shook my head negatively, the young man abruptly stood to his feet. He then grabbed his bedroll from the sidewalk, and without another word strode off into the night. Not at all did I regret losing the rest of that interview.

Especially helpful in overcoming engulfment—in this instance of being so overwhelmed that I would abandon the research—were the visits with relatives in California and Minnesota that I had sandwiched into my itinerary. Although I had planned those visits primarily for convenience, they proved to be a refuge from the harsh, threatening world of the homeless. Escape hatches into the "normal," those respites mitigated the impact of this overpowering world of the streets, allowing me to "get my bearings" so I could continue the research. There I was able to spend time with people whose world was like mine, whose assumptions of reality did not impinge upon my reality orientations. There people were embedded in family units, people whose routines included regular hours for sleep and regular hours for work, people who ate meals at small tables in a room called a kitchen, people who sat in what they called "living rooms" working out a common reality by watching television and discussing what had happened at work and school (Berger and Kellner 1964).

To experience the unknown, the uncertain, and the novel, can be, of course, major attractions of field work. But to see people locked into such degrading and despairing poverty, or to see a youth attack an older man who then must be carried off to the hospital, and to have to always be suspicious of those around you—such things exact an emotional toll. Without exaggeration, it was not until three months after I returned home that I was able to sleep a full night without awakening at least once.

SOCIO-BIOLOGICAL CHARACTERISTICS AND FIELD RESEARCH

As Gustav Ichheiser (1970) used to emphasize, in order for social science to advance we must make visible the taken-for-granted aspects of life in society. If we do not, we will overlook their relevance and unwittingly bootleg those background assumptions into our conclusions about social behavior. It is no less necessary to explicate the taken-for-granted assumptions that underlie the very doing of our sociological research. If those assumptions remain unexamined, they unwittingly may become incorporated into our findings. If unexamined, we cannot know how or to what extent they influenced those findings.

Gender, race, and age are among those assumptions. As master traits, gender, race, and age are the major identifiers in social life. Rather than superficial characteristics, they are essential parts of everything we do, chief determinants of our interaction and communication. In a myriad of ways, many or perhaps most of which lie below our level of consciousness, those identifiers influence our perspectives, helping to determine what we see and how we feel.

Field research can be viewed as a form of communication and interaction. As such, it is affected by whatever factors help determine the form and content of communication and interaction. Consequently, it should be obvious that the gender, race, and age of the researcher, through the meanings they communicate, can influence significantly the research process and thereby the conclusions we draw from our observations. Yet these socio-biological characteristics continue to be taken-for-granted assumptions, remaining essentially unexamined as major factors in social research.[8]

Gender

As in other field studies, the intertwining of gender and sex roles was significant in this research. For example, these research settings would have posed greater risk for myself if I were female, especially in terms of vulnerability to rape (cf. Adler 1985, p. 27). While I do not look especially "tough" (or not even tough at all, I suppose), simply being male afforded some degree of protection.[9]

In addition, because the homeless are overwhelmingly male, being male allowed entry to aspects of the homeless experience that I otherwise would have been denied. On the obvious level are areas of sleeping and showering, to which a female researcher would have been denied access by shelter personnel, or if somehow gaining such access, in such vulnerable settings likely would prove prey to individuals or groups or marauding males. Similarly, many of my interviews were conducted after dark, some in relatively isolated settings.

For example, I can recall one night's search for the homeless on the streets of New Orleans that ended up in Lafayette Square. There I found homeless men who, in spite of the evident presence of scurrying rats, preferred sleeping in this little park to sleeping in the filthy shelter only a few blocks away. Although this setting proved productive, yielding interviews with more independent-minded homeless, it also may have posed greater risk to a lone female researcher.

Being male also proved a hindrance in this research as it denied me access to many aspects of the female experience of homelessness. Although I was able to interview homeless females, and to gain a variety of such respondents in terms of age, race, length of homelessness, and degree of isolation, my own gender is inhibitive. For example, my background of experiences leads me to interpret the world from some common connection with males. Although my experiences are unique, and vastly differ from the homeless males whom I interviewed, there was yet a bond of maleness, a common connection that is most difficult to explicate, but that nonetheless was present. This very mutuality, so to speak, provided a common basis for an interpretative understanding of their events and experiences.

So it is with female researchers and female subjects. A common bond due to similarities of experiences, a jointedness flowing from *sitz-im-leben,* molds understandings, allowing for a form of Weberian Verstehen that changes with the gender of the researcher. By no means does this signify that I cannot understand what homeless females told me about their experiences of becoming and being homeless, nor that I am unable to relate to readers essential meanings of those experiences. Rather, there likely are many things to which a female researcher would have been more sensitive to than I, and thus may have explored during the interviews that my own gender and sex role background impeded and left invisible. Equally as significant, from identical interviews I have less *Verstehen* when it comes to grasping a shared meaning of how it feels to be homeless *and* female.[10]

Interestingly, my gender did not stop me from being able to stay at a shelter for women (cf. Shaffir, Stebbins, and Turowetz 1980, pp. 13-14). This overnight visit, however, was due to one of those serendipitous events that affects social research. During my preliminary visit to Washington, DC, I had interviewed the male member of a husband and wife team who operate multiple facilities for the homeless. When I recontacted them during this phase of the research, due to their confidence in who I was and their desire to see that I expand my data base, they invited me to spend the night in their shelter for women. This experience allowed me to observe the contrastive workings of this shelter, note differing orientations on the part of female homeless, and obtain interviews that otherwise would have remained inaccessible. My gender, however, prevented me from participating in some areas of this shelter, especially those that had to do with the bathroom, dressing, and undressing.[11]

There remains, finally, a matter of gender that is at least equal in significance to those of access and interpretation. When it comes to gender, to what are the subjects responding? Assuming that one has equal access, in what ways would one's data be different if one's gender were different? I think that it is imperative to assume that both form and content are involved, that is, that subjects will both express themselves differently to a researcher on the basis of the researcher's gender and also change the actual content of what they express. Thus males will relate certain experiences to a male researcher that they will not relate to a female researcher, and other experiences to a female researcher that they will not recount to a male researcher. It is the same with females.

One can only attempt to indicate the significance of gender in social research. It would take a full-fledged experiment using matched samples and comparative interviewing by gender to begin to uncover the dimensions that this variable plays. And I am convinced that the results of such an experiment would be suggestive only, merely unearthing indicators of such differences.

Race

Although the race of the researcher inexorably affects research, its role in the research process and in research findings also remains underexamined. (For allusions to its significance see Johnson 1975; Kleinman 1980, p. 179; Mann 1970, p. 120; Posner 1980, p. 208; and Wax 1971, p. 46, while for a more detailed, self-reflective evaluation of its role, see Liebow 1967, pp. 248-251.) Certainly being white was not a hindrance in this research. Most shelter personnel were white, and this correspondence of race between researcher and researched probably made them more comfortable in my presence, increased their rapport, and maximized their cooperation. I assume, indeed, that whites who were racially prejudiced were aided by my being white. No black, Indian, or Chicano homeless person, however, refused to talk to me for which I could ascertain or even surmise race as the basis.[12] They were as willing—or as unwilling, as the case may be—to talk as were the whites.

Similar to the instance of gender, however, it is likely, and some would say inevitable, that there were things that blacks would have said, or would not have said, or would have said differently, if I, too, had been black. It is the same with Indians or Chicanos. I am certain indeed that this "covering" effect of race is there, but I find myself unable even to speculate regarding the specifics of its significance.

Age

Age is the third socio-biological characteristic that plays a significant role in social research. Indeed, of the various researches in which I have been involved, this was the first time that I was even aware of the significance of age in the research process. Prior to this, age had been below my level of awareness: It had simply "been there," and, supposedly at least, irrelevantly present.

At this point in my life, however, something was markedly different: I found that I had aged since I had last been in the field![13]

When I did participant observation research on cab drivers, I was about 29 (Henslin 1967). Indeed, when I think about it, I was a fairly young (or naive, if you prefer) 29 at that. I gave little thought to danger, as I was caught up in the excitement of the sociological pursuit. Although two or three cabbies were stabbed the first week that I drove a cab, certain that such a thing would not happen to me, I gave the matter little thought.

In the ensuing years I had retreated to my sanctuary, that little place of safety and refuge known as the professor's office. There I conducted most of my research, in private talking to people about their experiences with a variety of deviances. Untouched, seldom moved, from this position of aloofness I gathered information on a variety of topics and behaviors of sociological interest to satisfy my various curiosities (perhaps more respectably termed, the sociological quest).

Now, however, I was once again face to face with street realities, and at this point in my life things no longer looked the same. Age had accomplished what it is rumored to accomplish: It had brought with it a more conservative (in the literal sense of the word—conserving, drawing in, protective—in this case, of my own life) approach to street experiences. I found myself more frequently questioning what I was doing, and even whether I should do it.

This did not stop me from entering some questionable situations that I felt were necessary for the research, but it did others. For example, when I realized that runaways were missing from my sample, I made a trip to Hollywood, California, to seek them out. Although I searched for several hours, I had run down one blind alley after another and I was unable to locate runaways. They certainly were not advertising their status, and those whom I asked about runaways gave me a rather wide array of responses, most of which were varieties of indicating that it was none of my business.

When I finally turned up a more promising lead, I very much wished to follow it. About ten o'clock that night I turned the corner to a side street where I had been told that runaways "hang out." Down the block I saw about half a dozen or so young males and two females clustered in front of a parking lot. Somehow they did not look like the midwestern suburban youth I had come to know. What was most striking about this group was the amount of "metal" they were displaying, notably the studs protruding from various parts of their bodies.

A few years back those youths would have struck me as another variant group that likely had engrossing experiences to relate. No longer. They now impressed me as a group that discretion would indicate as being better off left alone. I broke my stride momentarily and rapidly surveyed the setting. Noting that a nearby streetlight illuminated the scene and that it was only a half block off the frenetic activities of Hollywood Boulevard, I concluded that I could beat a hasty retreat if one became necessary. I then terminated my momentary pause and continued toward the youths.

I did get the interviews with runaways, and they were among the best I obtained in this research. However, to accomplish this I had to overcome the greater caution that increasing age had brought.

And there were times that I did not overcome it. For example, in the above situation the two runaways I interviewed told me that they slept in abandoned buildings. I immediately began to wonder what this was like: How did they protect themselves? How many slept in the same building? Was it like the imagery from *Oliver Twist* that immediately came to mind? How did they choose their buildings? Where did they keep their clothing and personal items? The two females whom I was interviewing were clean and well groomed. How did they manage this, sleeping in abandoned buildings?

I very much wanted to learn the answers to those questions, but that desire made me highly aware of my age. Had I still been in my twenties, I probably would have invited myself along. The rapport with those runaways was excellent, and I think that I would have been welcome to enter their world. But at age 47 an immediate rebuttal came to mind, and it centered on two things. The first thing I heard resounding in my head was, "Safety, safety, safety." This was accompanied by quickly-played scenarios of dangers, of "What ifs . . ."

Now age and other background variables cannot be separated from social class, personality, marital status, attitudes toward marriage, religious orientations, and the like. At this point, all of these factors also came into play. In addition to my social class background of experiences causing me to feel that an abandoned building would not be the most desirable place to spend the night, I also felt that my wife would not approve of such sleeping arrangements, and that the situation may contain temptations that would conflict with my spiritual orientation. Consequently, the second thing that I heard resounding in my head was, "What would it look like, you doing that?"

All of this took only fleeting seconds, of course, and occurred during the interview itself. Before I was able to resolve my mental dilemma, or, more accurately, as a means of resolving it, the salience of my transportation came to mind. I had been taken to Hollywood by two persons who were going to leave in about half an hour. Most of the evening had been in pursuit of the runaways, and there now remained only enough time to complete the interview before my transporters would be leaving.

CONCLUSION: EMPIRICISM, THEORY, AND PRACTICALITY

Of what value is such a study? Is such research to be measured only in terms of how it satisifies the intellectual curiosity of the researcher?

I think not, although one can make the argument of empiricism for empiricism's sake. It seems to me that the value of this research, as well as all field research, and perhaps all research regardless of method, properly ought to be measured along three dimensions: its contributions to empirical findings, to theory, and to social change.

Although it is too early even to attempt to estimate the value of this research along any of these dimensions, the data being only in preliminary analysis and far from published form, I am hopeful that it will contribute on the first criterion. I do have data on what kind of people are "really" "out there", how they got there in the first place, and what their life now is like. For example, I have done a preliminary analysis of the types of homeless on the streets of America, one that matches the realities of the people on the streets, not some stereotypical perception of who is "out there" (Henslin 1985).

These empirical findings also hold the potential of contributing modestly to theory, at least in the sense of connecting the different types of homeless to their various routes to homelessness. This may help us to see similarities and differences in the routes that the distinct types of the homeless have traveled. Additionally, as the analysis proceeds, the potential for contributing to sociological theory exists on both the macro and micro levels. On the macro level is the potential of tracing the empirical findings of routes to homelessness to social structural factors, such as the individual's original location in social class, family, and community. On the micro level lies the potential that comes from examining the individual's role in the route to homelessness, the coping mechanisms and mapping strategies now utilized by these people, and abstracting therefrom principles that apply to a broad range of situations.[14]

Finally, there is the matter of practicality. This research focuses on a social problem of severe dimensions, both for the society experiencing homelessness in its midst, and for the individuals who find themselves homeless. In the midst of such social disruptions and individual suffering, it seems to me the height of intellectual insensitivity to conduct studies that, though empirically grounded and theoretically sound, contribute nothing to relieve the suffering on which they are based.

Those are real people, and their despair and pain are similarly real. Although we are not social workers, and we do not want to become such, if our findings and interpretations hold significance for alleviating at least somewhat that despair and pain, then we need to help make those findings and interpretations applicable to social change. Note that I did not suggest that sociological findings and interpretations might transform society. I hold no such illusions about the ultimate value of sociology, but certainly our findings, as well as our theoretical interpretations of those findings, hold the potential of at least "alleviating somewhat" the suffering of this present world, and in my view, we owe it to our subjects, and to ourselves, to purposefully make such research results and conclusions available for such purposes.

Nor do I hold illusions about the theoretical or practical, much less any ultimate, value of this particular research. It may end up having been worthwhile only in terms of the satisfactions it provided my curiosities about the worlds of the homeless. Or it may contribute also on the empirical, theoretical, and practical levels. At this point, I do not know—but that also is one of the fascinating aspects of social research, to anticipate, but not to know.

ACKNOWLEDGMENT

I wish to express my appreciation for the grants from the Graduate School of Southern Illinois University, Edwardsville, which helped me to conduct this research.

NOTES

1. Writing this makes it sound as though I had not considered such potential conflict. That would be far from the truth as my wife and I held many detailed discussions on this topic. Fortunately, she has held her tongue (mostly), avoiding the "I-told-you-sos" that she could use quite accurately.

2. Before the field work began, I had envisioned the results as including a series of photographs of the homeless across America. For this purpose I purchased a Canon camera with automatic features, extra lenses, and the like. It quickly became apparent to me that it would be unwise to use this camera, or even to bring it with me to skid row.

3. This choice of attire brought with it an unanticipated consequence. Since my everyday clothing was similar to the everyday clothing of the homeless, it helped me realize that in spite of the many other life differences the "gap" between myself and the homeless is much less than I had thought.

4. On the airplane I also carried an overnight shoulder bag in which I stashed my extra recording tapes, taped interviews, airplane ticket, traveler's checks, and extra clothing. I routinely stored this piece of baggage in a locker at the airport or at a bus station on skid row.

5. Actually there is such a map, although it is long outdated. In his 1963 work, Bogue provides maps of skid rows of 41 of America's cities.

6. Although I am using the subjectively active verb "to gather" to refer to the collecting of deviants in this area, it is more accurate to state that deviants are segregated into this specific locale. In some instances, city officials use zoning laws to force particular deviants into specific areas, such as sometimes is the case with pornography. Most segregating of deviants, however, is much more informal than this. Deviants are aware that laws are enforced more strictly or casually by the area of the city, that in some neighborhoods they will be harassed by officials, while in others they can remain relatively unbothered. Highly visible, outcast alcoholics are a traditional example (cf. Bittner 1967; Chambliss 1964; Giffen 1966; Hobfoll, Kelso, and Peterson 1980).

For the homeless, both formal and informal controls similarly operate: Although zoning per se does not limit the homeless to particular urban areas, city officials accomplish the same purpose by requiring permits to operate shelters, thereby keeping those shelters out of areas whose residents have more political power or who are more likely to complain.

An example of more direct social control of American homeless is the attempt to zone the homeless out of an entire city. The city officials of Phoenix, Arizona, tried to remove the support systems of the homeless by closing the city's shelters, alcoholic dry docks, and residential hotels (Higgins 1983; *The Economist* 1982)

7. The emotional problems sustained by the researcher due to the discordance between the researcher's experiences with the homeless and the middle class have also been mentioned by Estroff (1981, p. 4).

8. For a sampling of the mentioning of the significance of background characteristics in field research, see Daniels (1967, 1985); Prus (1980); Shaffir, Stebbins, and Turowetz (1980); and Wax (1971). For a fascinating analysis of how the subjects' perceptions of the researcher's "real" social world negatively affect the data-gathering process, see Wieder (1983). Background variables are salient not only for field research, of course, but for other types of social research as well (cf. Friedman 1967; Rosenthal 1966; and Rubin 1977).

Background variables that I hope to one day see analyzed by field workers include social class, body type, height, health, sexual orientation, attractability, personality, marital status, attitudes, and values—where there will be a double foci, their effects noted not only on those being researched, but also on those doing the researching.

9. Jacqueline Wiseman's study of skid row alcoholics is ample evidence that gender need not be a barrier in this setting. The skid rows of American cities, however, have changed markedly since the 1960s when Wiseman did her research, with the camaraderie of drinking companions largely giving way to severe risks posed by the presence of large numbers of alienated, angry young men. In addition, Wiseman did not sleep in the skid row shelters.

10. This brief mentioning of the role of gender in *Verstehen* hardly does justice to its complexities, and I hope one day to see its in-depth treatment. For example, what are the bases for greater *Verstehen* with some subjects rather than others even though the sex of the researcher and the subjects are the same? Or what is the role of atypicality of gender experience and *Verstehen* by sex of subject?

11. One of the few researchers to stress the significance of gender is another reseracher who worked with the homeless. Sue Estroff (1981, p. xviii) says: ". . . being female helped and hurt. Over half of the subjects were men. My gender served as an entree to contacting them and eliciting some interest, but it created tensions as well. Many had never had a female friend, that is, a symmetrical, platonic, heterosexual relationship. This led to some confusion on their part when their sexual advances offended me, and to reluctance on my part in entering situations with them that might be misconstrued. It was often inappropriate to participate with the group as the only female, and as a sexually inaccessible one at that. In a world where nurses and other female staff were also sexually off limits, this problem undoubtedly maintained some barriers to inside communication. Without question, a male co-worker could have provided a different type of observation and information."

12. I have found it fascinating that so far I have not come across any Oriental homeless. I am certain they must exist, but if they do they either are so few that our paths have not yet crossed or, if they exist in anything close to their proportion of the population, they are somehow spatially segregated.

13. I am referring here only to long-term projects involving street life. On occasion I have made short excursions into the field, from staying with hippies on Haight-Ashbury to transvestites in Dupont Circle in Washington, DC, while longer-term, non-street field research has taken me to various areas of Mexico, Europe, and Africa.

14. For an example of this latter point based upon my earlier field research, see Henslin (1969).

REFERENCES

Adler, P.A.

1985 *Wheeling and Dealing: An Ethnography of an Upper-Level Drug Dealing and Smuggling Community.* New York: Columbia University Press.

Anonymous

1982 "In the Bleak of Winter." *The Economist* 285(December 25):43, 46-47.

Anonymous

1984 *On the Streets: Report of the Mayor's Task Force on Street People.* Lexington, Kentucky.

Arce, A.A., Tadlock, M., Vergare, M.J., and Shapiro, S.H.

1983 "A Psychiatric Profile of Street People Admitted to an Emergency Shelter." *Hospital and Community Psychiatry* 34: 812-817.

Berger, P., and Kellner, H.

196 "Marriage and the Construction of Reality." *Diogenes* 46:1-23.

Bishop, K.P.

1983 *Soup Lines and Food Baskets: A Survey of Increased Participation in Emergency Food Programs.* Washington, DC: The Center on Budget and Policy Priorities.

Bittner, E.

1967 "The Police on Skid-Row: A Study of Peace Keeping." *American Sociological Review* 32: 699-715.

Bogue, D.J.

1963 *Skid Row in American Cities.* Chicago: Community and Family Study Center.

Chagnon, N.A.

1983 *The Yanomamo: The Fierce People,* 3rd ed. New York: Holt, Rinehart, and Winston.

Chambliss, W.J.

1964 "A Sociological Analysis of the Law on Vagrancy." *Social Problems* 12:67-77.

Cimons, M.

1976 "A Refuge for Homeless Women." *Los Angeles Times,* May 23, pp. 1-4.

Clark, M.C.

1981 "Los Angeles: Burgeoning Mixed-use in Downtown." *Urban Land* 40:19-22.

Coleman, J.R.

1988 "Diary of a Homeless Man." Pp. 171-184 in J.M. Henslin (ed.), *Down to Earth Sociology: Introductory Readings,* 5th ed. New York: The Free Press.

Crystal, S., Goldstein, M., and Levitt, R.

1982 *Chronic and Situational Dependency: Long Term Residents in a Shelter for Men.* New York: Human Resources Administration of the City of New York.

Daniels, A.K.

1967 "The Low-caste Stranger in Social Research." Pp. 267-296 in Gideon Sjoberg (ed.), *Ethics, Politics, and Social Research.* Cambridge, MA: Schenkman.

Daniels, A.K.

1985 "Engagement and Ethical Responsibility in Field Work." *SSP Newsletter* 16:12-14.

Estroff, S.E.

1981 *Making It Crazy: An Ethnography of Psychiatric Clients in an American Community.* Berkeley: University of California Press.

Friedman, N.

1967 *The Social Nature of Psychological Research: The Psychological Experiment as a Social Interaction.* New York: Basic Books.

Garfinkel, H.

1956 "Conditions of Successful Degradation Ceremonies." *American Journal of Sociology* 61:420-424.

Giffen, P.J.
1966 "The Revolving Door: A Functional Interpretation." *Canadian Review of Sociology and Anthropology* 3:154-166.
Goffman, E.
1959 *The Presentation of Self in Everyday Life.* New York: Doubleday.
1961 *Asylums.* Garden City, NY: Doubleday-Anchor.
Henslin, J.M.
1967 "Craps and Magic." *American Journal of Sociology* 73:316-330.
1969 "Guilt and Guilt Neutralization: Techniques of Adjustment in Response to Suicide." In J.D. Douglas (ed.), *Deviance and Responsibility: The Social Construction of Moral Meanings.* New York: Basic Books.
1985 *Today's Homeless.* Paper presented at the 35th annual meeting of the Society for the Study of Social Problems, Washington, DC.
Higgins, M.
1980 "The Anchorage Skid Row." *Journal of Studies on Alcohol* 41:94-99.
1983 "Tent City." *Commonweal* 110(September 23):494-496.
Ichheiser, G.
1970 *Appearances and Realities.* San Francisco: Jossey-Bass.
Johnson, J.M.
1975 *Doing Field Research.* New York: Free Press.
Kleinman, S.
1980 "Learning the Ropes as Fieldwork Analysis." Pp. 171-183 in W.B. Shaffir, R.A. Stebbins, and A. Turowetz (eds.), *Fieldwork Experiences: Qualitative Approaches to Social Research.* New York: St. Martin's Press.
Liebow, E.
1967 *Tally's Corner: A Study of Negro Streetcorner Men.* Boston: Little Brown.
Mann, F.C.
1970 "Human Relations Skills in Social Research." Pp. 119-132 in W.J. Filstead (ed.), *Qualitative Methodology: Firsthand Involvement with the Social World.* Chicago: Markham Publishing Company.
McCall, M.
1980 "Who and Where are the Artists?" Pp. 145-158 in W.B. Shaffir, R.A. Stebbins, and A. Turowetz (eds.), *Fieldwork Experience: Qualitative Approaches to Social Research.* New York: St. Martins.
Pollner, M. and Emerson, R.M.
1983 "The Dynamics of Inclusion and Distance in Fieldwork Relations." Pp. 235-252 in *Contemporary Field Research: A Collection of Readings.* Boston: Little Brown.
Posner, J.
1980 "Urban Anthropology: Fieldwork in Semifamiliar Settings." Pp. 203-212 in W.B. Shaffir, R.A. Stebbins, and A. Turowetz (eds.), *Fieldwork Experiences: Qualitative Approaches to Social Research.* New York: St. Martin's Press.
Powdermaker, H.
1966 *Stranger and Friend: The Way of an Anthropologist.* New York: W. W. Norton.
Prus, R.
1980 "Sociologist as Hustler: The Dynamics of Acquiring Information." Pp. 132-145 in W.B. Shaffir, R.A. Stebbins, and A. Turowetz (eds.), *Fieldwork Experiences: Qualitative Approaches to Social Research.* New York: St. Martin's Press.
Rosenthal, R.J.
1966 *Experimenter Effects in Behavioral Research.* New York: Appleton-Century-Crofts.
Rubin, Z.
1977 "The Love Research." *Human Behavior* (February).

Shaffir, W.B., Stebbins, R.A., and Turowetz, A. (eds.)
1980 *Fieldwork Experiences: Qualitative Approaches to Social Research.* New York: St. Martin's Press.
Strasser, J.A.
1976 "Urban Transient Women." *American Journal of Nursing* 78(December):2076-2079.
Thompson, W.E.
1988 "Motorcycle Gangs." Pp. 161-170 in J.M. Henslin (ed.), *Down to Earth Sociology: Introductory Readings,* 5th ed. New York: The Free Press.
Wax, R.
1971 *Doing Fieldwork: Warnings and Advice.* Chicago: University of Chicago Press.
Wieder, D.L.
1983 "Telling the Convict Code." Pp. 78-90 in R.M. Emerson (ed.), *Contemporary Field Research: A Collection of Readings.* Boston: Little Brown.
Wiseman, J.
1970 *Stations of the Lost.* Chicago: University of Chicago Press.

EXPECTATIONS AND REVELATIONS:
EXAMINING CONFLICT IN THE ANDES

Helen Rainbird

INTRODUCTION

A central problem confronting the researcher analyzing class struggle, both contemporary and historical, concerns the interpretation of documentary and oral sources of information provided by participants in the struggle. These accounts may contradict each other and so the substantiation of data is, therefore, a major concern. In addition, the researcher has to cultivate a certain appreciation of these sources of information as *interpretations* of these events concealing more complex social processess, which may be only partially understood by the participants or deliberately manipulated by them for one reason or another. The researcher also interprets, but has the disadvantage of carrying a number of preconceptions about the nature of political processes. In this account, these preconceptions concerned the role of clandestine and illegal activity and its relationship to commonly accepted and documented versions of events.

Studies in Qualitative Methodology, Volume 2, pages 77-98.

ISBN: 1-55938-023-3

In this chapter I investigate some of the issues arising from studying political conflict and, in particular, the cultural and perceptual preconceptions that Europeans carry with them when they attempt to analyze political processes and organizations in the Third World. The fieldwork experience I discuss is a period of twenty months spent in Peru between Feburary 1975 and September 1976 during which I studied the implementation of the 1969 Agrarian Reform. This necessarily required an analysis of the historical events both at national and local levels that led up to the passing of the Agrarian Reform and substantially affected its implementation in the fieldwork area.

I was particularly interested in studying the Agrarian Reform in light of debates on underdevelopment and the role of peasant organization in community politics. I chose a fortunate moment in Peruvian history to examine these issues since the 1969 Agrarian Reform introduced by the reformist military régime of General Juan Velasco was widely acclaimed as one of the most thorough and radical land reforms in Latin America and was in the final stages of implementation when I was in the field. It was an exciting time to be in Peru in the "first phase" of the military government before more rightwing tendencies were consolidated when General Morales Bermúdez assumed power in a palace coup in August 1975. There had been land reform, expropriations of the oil refineries and sugar plantations, experiments in workers' self-management in factories, a literacy campaign: the rhetoric of the Peruvian revolution led commentators such as Eric Hobsbawm to claim that this was a third path to socialism (Hobsbawm 1971) though the non-Communist Peruvian left did not share this view.

PREPARATION FOR FIELDWORK

My preparation for fieldwork had been an MA thesis on recent literature on the Peruvian peasantry but, with hindsight, I can say that no amount of theoretical or methodological study would have prepared me for the experience of participant observation in a remote Andean village. It was the real "rite of passage." The most valuable advice I received in psychological preparation for the isolation of fieldwork came from a fellow anthropologist, recently returned from Cusco, who told me to expect to get depressed every three months and to plan regular breaks away from the area. On establishing social relations with peasants he said I would make mistakes but people would forgive me.

Perhaps more importantly, none of the male anthropologists I knew was able to tell me what I was likely to meet as a single, European woman on my own in a *machista* (male chauvinist) society. Being female affected my relations in the field insofar as certain activities were exclusive to one sex or the other. Nevertheless, the fact that I towered over most peasants, wore trousers and was an outsider of high social status placed me in a rather ambiguous category that allowed me to attend meetings and visit people freely around the countryside as men did, but not to drink with the men unless other women were present (a major source of information, if our male colleagues are to be believed!) On the other hand, I had good access to women's activities and gossip networks, their warmth and affection.

Being female and single also made life outside my relations in the field more hazardous yet it had its compensations. As a woman with no father, husband or brothers to look after me I was considered to be easy game. Conversations with strangers on buses inevitably took the following course: "What is your name?" "Are you married?" "Tell me about 'free love' in your country." It was also considered inappropriate for a young woman to be staying on her own and I am eternally grateful to a number of families who took me under their wing. Several peasants offered me their daughters to keep me company and an agrarian reform official kindly arranged for me to be accompanied by his secretary to visit an expropriation he was overseeing in the next province, which involved an overnight stay. At the time I did not realize I was the pretext for them to have a night away together! My location outside local networks of kinship and affinal relations also made me appropriate as a confident both to a young single woman of my own age who was pregnant and to two older women who suspected their husbands were unfaithful. The problems associated with being single, a woman, and a foreigner were greatest in urban areas; I recall three attempted robberies and three attacks of a sexual nature not to mention the constant *piropos* (complements, often rather vulgar) and innuendos. In the countryside the situation was different because the people with whom I was living and working had literally just thrown off the yoke of serfdom. My social status allowed me access to areas of activity closed to peasant women. At the same time, the fact that I did not behave toward them in the way people of high social status normally did (i.e., I was polite to them rather than abusive, used the formal "Usted" rather than the familiar "tu" to them) made me an anomaly. I was invulnerable to physical attack but not to the wishful "perhaps you'll stay here and get married."

SELECTING AN AREA FOR STUDY

When I arrived in Peru in February 1975 with faltering Spanish and the addresses of a number of contacts, I had two major considerations. One was to find an area in which to do my fieldwork and which would enable me to examine my research topic. The other was to be not too far away from people with whom I could discuss my findings. When I discovered how bad communications were in Peru (landslides, floods, impassible roads, and so on), I decided against working in the Andes and settled instead on a field location on the coast, not too far from Lima where a number of friends and colleagues were working.

At this point, I should say that I arrived in Peru at the wrong time of year. When I arrived in Lima in February it was humid and hot. My first visits to the Andes made at the same time, were during the rainy season. My enduring first impression of Cajamarca was of being stuck on the road overnight on the high moorland behind a lorry that had got stuck in the mud on the unpaved road. My introduction to Bambamarca (where I eventually made my base) involved being covered in flea and bed bug bites and getting mud up to my knees during the hour-long walk up to the *casa hacienda* (the big house of the estate) of Chala. Communications between Chota, where all the legal and administrative offices were, and Bambamarca were infrequent and on my first visit I had to walk the twenty miles back to Bambamarca in the rain and the mud. I decided I could not work under such conditions and in such isolation. Therefore, when I was encouraged to look at the possibilities of working in the valley of Cañete, a couple of hours travel from Lima, it appeared to be an ideal solution: many migrants to the estates were from the Central Sierra (a particularly well researched part of the Andes), it was close enough to Lima to allow frequent contact with people with whom I could discuss my work, and I had introductions to people working and living in the area. I started working in Cañete.

Looking at the map of South America, Peru is close to the Equator and it could easily be assumed that this means it enjoys a warm climate. What I did not know was that in April the coast clouds over for nine months of the year with pea soup-coloured skies. It becomes cold and humid, washing goes mouldy when hung up to dry and, in my case, I caught infections of the skin, scalp, eyes, nose, and ears and continually suffered from the flu. The climate of the mountains in contrast transforms itself into that of a fine English summer's day. My ill health, combined with the realization that my theoretical preparation for studying the peasantry was inappropriate for an area of commercial cotton production and an agrarian structure that was altogether more complex from the point of view of class composition, forced me to abandon five months of documentary work, negotiation of access and preparatory fieldwork in Cañete and to reconsider Cajamarca. Good introductions to Bambamarca enabled me to make up quickly the time I had lost through my false start in Cañete.

Though at the time the switch in fieldwork location seemed a major disaster, on reflection the experience of starting work in an area of plantation agriculture was of great value conceptually. It meant that even in the comparative isolation of Bambamarca, I could relate what I saw around me to the wider economy and society of Peru. In particular, it enabled me to understand the links between the rural areas where peasant agriculture predominates and the more developed capitalist sectors of the economy, namely, mining, manufacturing and plantation agriculture. Having seen the other end of the wage labor migration chain, I appreciated the social and economic links between rural and urban areas. It was extremely useful for understanding theoretically the relationship between an estate system that appeared to have been organized along feudal lines (that is to say, it used servile labor and renting contracts rather than wage labor) and the fact that it had existed in an area of intense out-migration. That is to say, it posed the theoretical problem of how an institution based servile labor relations had continued to exist into the 1960s when all the preconditions for its dissoultion were present.

THE FIELD AREA

The former estate (*hacienda*) that I eventually worked on was situated near the small market town of Bambamarca, some 90 kilometers away from the departmental capital of Cajamarca (see Map 1). Communications were poor, and a boring, bone-rattling journey across the *jalca* (high moorland over 4,000 meters) to Cajamarca lasted eight hours in good weather and longer in the rainy season. The estate had been particularly conflictive from the 1950s onward, in a region in which mobilizations against landlordism had been relatively few. This relative quiescence has generally been ascribed to two factors. First, the *hacienda* Llaucán, which borders Chala to the South-East, was the site of "the most horrendous hecatomb" in 1914 and 150 peasants, including women and children, were massacred as a result of an uprising against increased rents (Alegría 1973, p. 385). Second, two major tendencies had been affecting the estate system in the region. These were, on the one hand, the division and sale of estates to peasant tenants and, on the other, the development of commercial agriculture in areas closest to coastal markets. Where commercial agriculture was introduced, this generally occurred through the sale of peripheral land to peasant tenants and through the recruitment of a wage labor force. Chala was, therefore, somewhat distinctive insofar as high levels of social control had been maintained over the peasantry until the 1960s. The estate had been known throughout the region as a "problem" estate and as a result of this history the relative success or failure of the implementation of the agrarian reform here was seen as crucial to the progress of the reform throughout the region.

HISTORICAL CONTEXT

Throughout the 1960s and, indeed, throughout history, struggles have occurred between peasants and landlords over access to land and the payment of rent. What distinguished many of the struggles in the Peruvian highlands in the 1960s from those of the preceding decades was that they were led or influenced by

Map 1. Bambamarca, Location in Department of Cajamarca

KEY

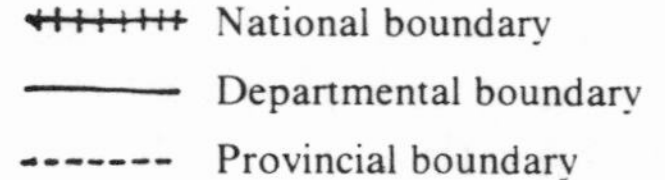

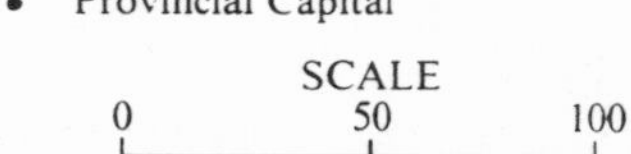

members of modern political parties rather than being "pre-political" struggles, concerned only with the narrow social world of the village pump (cf. Samaniego 1978; Quijano 1967). While some of these struggles were influenced by populist political parties such as APRA (the Popular Alliance for the American Revolution), others were developed under the influence of new political organizations whose strategy and tactics drew on the experience of the Cuban revolution. Alongside local level mobilizations against unpaid labor obligations and claims to ancient communal lands usurped by landlords, came demands for agrarian reform. Since the 1920s, a series of reforms had been proposed, but had never been successfully implemented due to the continued power of the landlord class and its allies. The military intervention of 1968 has been viewed as the response to a crisis within the dominant class, whereby industrial interests were unable to gain political control of the state and introduce reforms conducive to the reorganization of the economic structure. One of the crucial areas for reform was the landholding structure, which was perceived as a barrier to the development of industry and commercial agriculture. It was also a major cause of rural unrest, and threatened to undermine the first Belaúnde régime's (1963-1968) ability to control the dominated classes (Ferner 1983).

It is worth briefly documenting the development of the struggle in Chala since this informs the social backdrop to the theoretical issues I was interested in as well as affecting the conditions under which fieldwork was carried out. Although documentary sources were excellent (four volumes of legal papers relating to the expropriation that took over a month to get through), there are always problems of interpretation especially where the detective does not necessarily recognize the clues. Just establishing the events of the previous twenty years was a major feat. The earliest mobilizations are least well documented and oral accounts are vague on dates and figures. The first mobilization occurred in the early 1950s and was probably a response to increased rents and labor obligations introduced by a new manager, who later became landlord. In 1956 there was a further mobilization that extended to all sectors of the estate. It was organized in response to rent rises, the reduction of subsistence plots and the halving of midday food rations. However, it was precipitated by the demand for the schooling of peasant children and the injury of peasants in the construction of a road from Bambamarca to the estate house. At this stage the peasants relied on a legalistic strategy, collecting signatures for a petition to take to the Ministries of Labor, Education and Agriculture in Lima. There is evidence that members of the APRA party advised on tactics and some links were made with unions on the sugar plantations of Cayaltí and Casagrande on the coast and neighboring estate of Pallán. However, in this early phase of the struggle no permanent organizational structure emerged. Leaders were arrested and 83 families were evicted from the estate in reprisal.

The significance of the second phase of the struggle between 1963-1969 lies in the fact that politically and organizationally the mobilization was more developed. Moreover, the balance of forces had changed nationally, to the detriment of the landowning classes. The whole of the highlands was alive with the promise of agrarian reform, while the success of the peasant unions in La Convención (near Cusco), encouraged peasant struggles everywhere. A partial Land Reform was passed that allowed estates to be expropriated in areas of extreme conflict. This was the period of the massive land invasions in the Central and Southern Sierra, and the opening of guerrilla fronts in several zones.

The major point of difference between the earlier and the later stage of struggle in Chala lay in the formation of a peasant union organized by the Movimiento de la Izquierda Revolucionaria (MIR), the Movement of the Revolutionary Left, an organization which developed out of Rebel APRA, a split from APRA. It was formed in the late 1950s along with other groups of the Peruvian "New Left" under the influence of the Cuban revolution. However, while the strategy and tactics of the union were developed by the MIR, it organized clandestinely. Peasants taking administrative positions in the union and representing it publicly were not crucially important to the organization and in offical documentation and in peasant accounts it appeared that the main political influence had been APRA. There is plenty of evidence of this. For example, a communication written by the landlord writes that a local Aprista "came to Bambamarca and taking advantage of the large gathering of peasants here for the market, hired a loudspeaker . . . and from the balcony of a house incited his large audience, by announcing that he was Secretary of the Land Reform and that his party (the Aprista party) was making the agrarian programme effective" (Land Judiciary File, April 1965). The Aprista Deputy, Víctor Tantalean Vanini, was instrumental in trying to obtain an expropriation of Chala under the provisions of the partial Agrarian Reform Law, passed in 1964, which allowed for estates to be expropriated in zones of exceptional rural unrest.

In the earlier mobilizations, formal, legal channels of action had been used. This resulted in the peasants placing confidence in lawyers and in the Ministries in Lima. Though these channels of action were still used and the lobbying of sympathetic deputies in the Chamber of Deputies continued, this was a complement to mass mobilizations and action. Emphasis was continually laid on the peasants themselves taking control of the land, by subverting the effectiveness of the *hacienda* administration, in ways that led eventually to a situation in which the peasants were in de facto control of the *hacienda.* This was part of a strategy of creating incipient forms of "dual power" whereby the authority of the landlord was challenged by the authority of the union at local level. It was a strategy pursued by the Trotskyist led unions in La Convención and Lares (Southern Peru) and by the MIR in opening up guerrilla fronts in the Central Highlands in the 1960s.[1] From the onset of this phase of mobilization the demand was raised for the land, raising the key issue of the control of the means of production, which takes the directly political form of challenging the landlord's power. Evidence that a situation of this kind existed on the estate is to be found in a report by the Sub-Prefect of Bambamarca, which states "the peasants . . . refuse to pay their rents, and the *hacienda* is now treated as though it were a community with no respect for private property" (February, 1967). Later that same year the landlord wrote to the Minister of Agriculture "I believe the approved expropriation will be made valid, or rather, it is already valid by virtue of the peasants' de facto occupation of the land produced by the systematic invasion of the lands of the *hacienda*" (November, 1967). The strategy of creating dual power creates a confrontation between the power of the peasants on the one hand, and the power of the landlord and the repressive apparatus of the state on the other. This did actually occur in Chala. In 1967 the estate had been earmarked for expropriation under the provisions of the partial land reform law passed in 1964, but in June 1969 the landlord took advantage of the change of the head of the *Policía de Investigaciones* (equivalent to the CID) in Cajamarca, to order the arrest of one of the peasant leaders. The police entered the estate by night and tried to arrest him but were routed by some eighty peasants, men and women, who forced them to run for cover in the countryside.[2] On June 24th, thirty rangers were sent into the estate to rescue the police and if it had not been for the strange coincidence of the passing of the Agrarian Reform Law that day, it is certain that the confrontation would have escalated.

The final stage of the struggle, after the passing of the Agrarian Reform involved the mobilization of the peasants against plans for a production cooperative. In Chala, the implementation of the Agrarian Reform had to take into account the fact that there had been powerful mobilization against the landlord and that the peasants had been in de facto control of the land for a number of years. Furthermore, the terms of expropriation involved the creation of a cooperative and not the division and individual ownership of land, for which the peasants had fought. In Chala, the agrarian reform officials were confronted with peasants who were well-organized, and it was precisely this unity that had to be broken down in order to impose the Government's cooperative policy. Furthermore, it was important to set a precedent in Chala for the rest of the region, for the success of the cooperative here had important ramifications for other "problem" estates in the department of Cajamarca notably La Pauca and Culquimarca, where the peasants were also rejecting cooperative plans. Chala became the test case of Cajamarca and officials were keen to impose cooperative plans, however unsuitable and disliked, for the sake of political expediency.[3]

The agrarian reform officials were thus faced with contradictory goals. On the one hand, they had to undermine peasant unity that had grown up around the struggle for the land but at the same time, they had to organize the peasants themselves.

The peasant union had been disbanded by this time. Its most politically conscious leaders had joined other struggles and a number of opportunist leaders remained, who were more interested in making their own fortunes than in maintaining the unity of the peasant organization. At this stage the peasant mobilization was unable to sustain a clear class perspective in the struggle against the cooperative. Competition between leaders and their followings, and the formation of factions occurred. While the formation of factions by no means aided the task of the reform officials, there is evidence that they learned to manipulate it to achieve some of the political objectives of the Agrarian Reform.[4]

Many of the problems that the peasant organization experienced at this stage can be related to the fact that they had organized clandestinely and that there had been a limited mass development of class consciousness. For while all the peasants had clearly perceived their class interests as opposed to the landlord's they were ill-equipped to evaluate and organize in this new situation. The personal ambitions of a few leaders, combined with their control over the division of the land, meant that patterns of personalism and patron-client relations that had existed between landlord and peasants were established between peasants and peasants.

Due to peasant resistance, plans for a production cooperative were eventually abandoned, as were plans for a cooperative combining a core of collective production with individual household production. Eventually a compromise was reached whereby *grupos campesinos* (peasant groups), organized in each of the five sectors of the estate, received collective legal title to the land. In practice, this was a fiction that scarcely concealed the fact that the land had already been divided up into individual plots. There were some small-scale collective enterprises on what remained of the former desmesne lands, the majority of which had been invaded. However, these were voluntary organizations. A project for a eucalyptus plantation had been established (which, having no short-term benefits to the peasants, was scarcely the best choice to demonstrate the benefits of cooperation) and continued to be a source of conflict during the period of fieldwork.

The fact that peasants in the five sectors of the estate had been organized into putative peasant organizations called *grupos campesinos* to receive title to the land shortly before I arrived on the scene was undoubtedly one of the reasons I was directed toward working in this area. Had they not been successfully "organized" my entry into the area by means of official Ministry of Agriculture introductions would not have been possible.

RESEARCH PROBLEMS AND METHODS

One of the theoretical issues I wanted to explore was the debate on the differentiation of the peasantry, a debate that is concerned with the capitalist transformation of agriculture. In the context of Cajamarca in the 1970s, this inevitably meant examining the dissolution of the estate system and its impact on rural class structure. Lenin's classical formulation of the issue suggests that capitalism develops in the same way in agriculture as it does in industry. Consequently, the peasantry differentiates into rich capitalist farmers on the one hand, and a poor peasantry—the rural proletariat, on the other, in a process parallel to the development of the bourgeoisie and the proletariat in industry. The independent or "middle" peasantry are a feudal element that will eventually disappear (1972). Nevertheless, there is a qualitative difference between the development of capitalism in agriculture and industry, due to existence of a monopoly over the means of production in agriculture, the land, and the existence of rent that acts as a brake on the agricultural development. As a result, agrarian class structure does not mirror industrial class structure. Concentration of property does not automatically lead to the elimination of small peasant enterprises, but tendencies toward concentration and division can co-exist (Kautsky 1972, p. 189). Certainly the continued existence of peasant smallholding is not precluded by the development of capitalism and may in fact conceal proletarianization. Labor migration and legislation protecting the institutional structures of rural society may be extremely conducive to the maintenance of peasant smallholding (Samaniego 1978; Sulmont 1975).

The analyses of the internal differentiation of the peasantry would have required survey data on landholding, livestock, family size and structure as well as non-agricultural sources of income. However, at an early stage I rejected a questionnaire survey as a method of data collection. This was for a number of reasons. First, observation indicated that all peasants were poor by virtue of their small plots of land and limited access to cash income. The exceptions were a few leaders who had acquired larger plots of land and were said to be charging other peasants for access to land. The majority had a diet of maize and potatoes in which bread was a rare luxury. Cash income was derived mainly from the sale of produce or from casual labor paid at a rate of S.30 per day for adult males (equivalent to 30p at the time, which would buy 30 bread rolls). Eggs and milk were sold rather than consumed. Huts were made of adobe and stone with no running water, latrines, or electricity. Firewood and water for cooking had to be collected by women and children, the majority of whom walked barefoot. Infant mortality was high and most families could not afford to buy medicines when illness struck. It was estimated that 90% of the former tenants of the estate were illiterate. The estate system had effectively prevented the process of capital accumulation from occurring among the peasants though those who had paid cash rents had been involved in more market transactions than those who had paid labor rents. I, therefore, felt that it would be unlikely that I would be able to document a significant growth in social differentiation though this has doubtless occurred in the years since I did my fieldwork. A further consideration was that records on landownership were inaccurate and were not available over time. Population and land tenure figures that were available concealed a multitude of discrepancies, illegal sales of land, as well as absentee landlordism, sharecropping and renting that were not supposed to exist under the provisions of the expropriation. Second, since the estate was unlikely to yield much on social differentiation, as far as field data was concerned, my interest began to focus on conflicts over land and on people's perceptions of past and present problems. For this, a questionnaire survey was not necessary. Finally, an important consideration was the illiteracy of the vast majority of the population. This was due to the fact that the landlord had forbidden labor service tenants to send their children to school. It seemed to me that though a questionnaire survey is a good method for collecting data in a literate society, even though it may create a hierarchical relationship between researcher and research subject, this hierarchical relationship would be even more acute where the research subjects had only recently emerged from serfdom and could neither control nor understand what I would be writing down about them.

A further reason for rejecting a survey method was the political climate. I did not want to provoke any of the peasant leaders by asking awkward questions, which a survey would inevitably raise. These leaders could prevent my access to the area altogether if they wished. I was warned by villagers that the peasants on the former estate were *guapos* (fierce). Individually, they were far from being fierce, but in conceptual terms the depiction of the Chalinos collectively as *guapos* was extremely appropriate, due to developments in the struggle against the landlord in the 1960s. This historical experience resulted in a continuing perception that the Chalinos were a law unto themselves.

My instinctive feeling that participant observation rather than a survey method was the correct approach was confirmed by subsequent events. On one occasion in a general assembly I observed peasants lying to officials from the Ministry of Agriculture about the numbers of livestock they held. The numbers were required for the innocuous purpose of immunization, but in my field notes I recorded "one peasant remarked that when this had happened on previous occasions it had been to take their animals away". Their distrust of officials and continued insecurity of access to land (rumors that the landlords were coming back abounded after General Morales Bermúdez assumed the Presidency in a palace coup in August 1975) combined to make formal interviewing an impossibility.

Contacts in the Field

In the light of this insecurity and distrust of officials, it could well be asked, what chance did a foreign anthropologist have of winning people's trust and overcoming their suspicions? I can only answer on the basis of my own assessment. The impression I received of people's attitudes to me was that they were very curious and very friendly. As I walked along country paths I was constantly being bothered by inquisitive peasants who had no inhibitions in talking about their problems, especially in relation to the land. It took at least an hour to cross from one side of the village to the other due to the constant need to stop and converse. This contrasts markedly to reports I had received from anthropologists who have worked in Quechua-speaking areas of Peru and have found people dour and uncommunicative. I believe one reason for this is that my introductions into the area were exceptionally good. On the one hand, my official introductions through the Ministry of Agriculture had come through the one official who was not distrusted. He was referred to as "a good person, he didn't try to cheat us like the other officials." On the other hand, I had introductions through, and for a time lived in the same building as, members of the progressive Catholic Church. They also happened to be Europeans. Their identification with the peasants, and people's identification of me with them was extremely valuable. With the knowledge of hindsight I can also say that the families I met through the Church had a reputation of being *luchadores* (fighters) through their involvement in the struggle for their children's education in the 1950s but because they had been evicted from the estate they had not been involved in the later struggle when the peasant union was formed. Moreover, they were independent of the leaders who had been involved in the sale of land after the reform. They were not the only families I associated with, but it could be argued that this limited the development of other relationships. However, I am also convinced that my closer association with other families would not have yielded greater information about the involvement of the MIR in the development of the struggle in the 1960s. This is for two reasons. First, the condition of organizing clandestinely would have meant that not all peasants would have been fully aware of decision-making structures and strategy. Second, even if they had been known, I would not have been told about it.

Problems of Verification Arising from Illegality

Burton (1978, pp. 164-165) has written of the impossibility of using survey methods where people are "secretive, hostile, wary or underground." In his own fieldwork in a Catholic community in Northern Ireland, he was working with people among whom there was support for the activities of the Irish Republican Army. As indicated above, the people I encountered during my fieldwork were not hostile yet the fact that there had been clandestine political activity in the area resulted in my receiving an "official" or "public" version of events.

Throughout my period of fieldwork I had accepted the version of events I had been told about past struggles. This was because both written and oral accounts deriving from different interested parties coincided and I had no reason to suppose that circumstances had been otherwise. The story ran that the union was formed in 1963, a number of people were named as leaders (including an Aprista). There had been conflicts with the estate administration, blacklegs, the police, the landlord himself, not to mention trips to Lima and Cajamarca to bring pressure to bear on the authorities. Then after the passing of the Agrarian Reform problems began to emerge as some of the leaders started selling off the land and rivalries emerged between cliques in different sectors of the estate. Oral accounts and official accounts tended to coincide. This was because peasant leaders of this time were quite happy to inflate their own role in the struggle. It is also the version of events presented by the Peasant Federation of Cajamarca in publishing a series of oral histories of struggles in the region (FEDECC 1975). Documents from the land judiciary further indicated the importance of these leaders and the influence of APRA. All the evidence seemed to suggest a classic situation in which unity had been maintained as long as the struggle was clearly focused on the landlord, but once the aims of the struggle had been achieved peasant individualism reasserted itself and patterns of patron-client relations were re-established.

Shortly before I was due to leave the field I obtained additional information that completely overturned my understanding of all that had happened in the crucial years between 1963 and 1969. I was introduced to this informant by an unsolicited contact in Cajamarca, so in many respects I came across my most important item of information quite by chance. This informant told me of the invovlement of the MIR in the struggle in Chala, explained the strategy of creating incipient forms of dual power, and the problems of organizing clandestinely and maintaining security. This information left me feeling that social reality was impenetrable, because however long I had remained in the field I would not have obtained this information from other sources. This version of events was corroborated by two references in my documentary sources to which I had previously attributed little significance. It is consistent too with other literature that has been published on guerrilla organization (Alegría 1987; Cabezas 1987; Guevara 1968a, b). In all, it demonstrates how easily public versions of events can be presented to the social science observer and how counter versions may be deliberately concealed not just from the observer, but also from people participating more directly in events. However, in this instance, it was also the case that my lack of awareness of the significance of the references to the MIR in the documentary sources resulted in my failure to link this to an explanation and analysis of the socal processes that were being recorded in great detail in the legal and administrative documentation held by the land judiciary. Primarily it was a case of one model of political processes and organization based on my own experience in Britain being at variance with modes of political activity and organization in Peru.

On re-examining my documentary sources in the light of this knowledge, I found it easier to sift through contradictory evidence since this was part of the strategy of undermining the authority of the landlord. Moreover, the landlord, having been forced into a rearguard action by the developing strength of the mobilization, also had to undermine the legitimacy of claims made by the union and to ensure that he retained as much land as possible once the expropriation took place. At a certain stage of the struggle the peasant union claimed that the landlord was an absentee rentier, producing photographs of empty pastures and his falsified tax returns as evidence. The landlord at the same time claimed he had 2000 head of cattle pastured on the estate, and that this herd was to be the basis of a small dairy farm he hoped to continue to run on the desmesne lands after the expropriation. Later on, he denounced a series of peasant invasions to the police. Indeed, it has been common practice in the Andes for peasants claiming rights to land to construct houses overnight to give the appearance of lengthy occupation and certainly there were invasions of peasants returning to lands from which they had been evicted in previous years. In some instances this tactic was used to victimize peasant leaders, but the union also denounced blacklegs and informers to the police as invaders, thus creating considerable confusion as to the identity of legal and illegal occupants.

Problems relating to legality continued to exist at the time of my fieldwork, since there were a number of dealings in relation to the land which were formally circumscribed in both the letter and the spirit of the 1969 Agrarian Reform Law. These included land sales, absentee landlordism and share-cropping on the expropriated land of *hacienda.* Rather than being secretive about these issues, people were only too willing to discuss them, and freely talked about the leaders who were causing the problems. Frequently, too, disputes over land, and particularly disputes over access to the former desmesne lands of the estate, were aired publically at general assemblies of the *grupos.*

Finally, cocaine refining (*pichicato*) and opium cultivation occurred in the region but again, their extent was difficult to verify. Accusations of drugs trafficking had figured in both the landlord's accusations against some peasant leaders, and in the peasants' accusations against the landlord. Though I have no evidence on the former, indirect evidence for the latter is that the landlord's brother was convicted on a drugs trafficking charge in 1949. If, as the union claimed, narcotics had figured in the landlords' interests in maintaining the estate, this could well account for the intransigence shown in dealing with demands for schooling and abolition of labor services. However, this remains an open question.

The Researchers' Social Relations in the Field

Despite all the problems outlined above establishing social relationships in the field was remarkably easy. Not only were people curious to talk to me but I was welcomed into their homes. Even business transactions, such as buying straw for a mattress, developed into social relationships—an example of "pluri-functional relations" as Godelier (1977, p. 67) calls them, in societies in which kinship relations play a dominant role in social life. Trying to find pasture for my horse (a major headache) also brought me direct experience of problems over access to land. I needed it because the farthest parts of the estate were an eight hour walk away from town though most of my work was conducted within a two hour walk. This horse was an endless source of problems. It was lent to me by the parish, but I had to find pasture for it. There was nowhere suitable in the village, so I had to ask the *grupo campesino* if I could pasture it on the estate. They refused on the grounds that I would then lay claim to the land. In the end, one of the peasant leaders looked after it, two hours walk away from the village, which effectively meant I could not use it. This subsequently involved me in interminable debates with his neighbors and the parish about what was a suitable rate of payment. The neighbors thought he should not be paid at all because he had had the use of the horse while pasturing it. The parish thought he should be paid a substantial sum because he had cured some sores on its back. I eventually gave him a small radio in exchange (the neighbors thought this was excessive) and I was subsequently inundated with requests for radios, indicating the public nature of all transactions in close-knit social groups. While participant observation enables the researcher to experience as well as to observe social action, it is the depth of social relationships established with the research subjects that makes it so rewarding.

Considering that the people who befriended me had little material wealth and lived on a diet of maize and potatoes, their generosity was overwhelming. In the course of a visit, I would be offered two or three platefuls of *caldo verde* (a potato soup flavored with a green herb), *chochoca* (ground maize soup) or *cancha* (toasted maize). I was invited on barter trips, taken to parties and Saint's Day fiestas, told of my appearances in people's dreams and discussed with relish the delights of scratching where the fleas had bitten. But I was also told of illness, the death of children and asked to take a mentally retarded child to my country to cure her, a request I could not fulfill.

While some of the families I came to know were motivated by genuine friendship and no further interest in material gain than perhaps acquiring my boots or buying my paraffin lamp when I left the field, others, particularly the leaders, viewed me as a resource to be manipulated. Some saw me as a person whose relatively high status could be used to enhance their own. In one general assembly I was introduced as someone who could help the group with legal dealings in Cajamarca and Lima. In another sector my help was requested to sell a eucalyptus plantation to pay off the agrarian debt. Though I was happy to use my social status to help obtain medical treatment and books for the communal library (set up through adult literacy classes), I avoided any involvement with dealings on behalf of leadership cliques, since I wanted to avoid identification with particular factions, especially where it was believed that abuses of power were occurring.

Fortunately, I had been warned by the Ministry of Agriculture official mentioned earlier that the relationship between two leaders, in particular, was strained. These were two of the leaders who had been involved in the sale of land, the renting of former desmesne lands and the protection of absentee landlords. When I visited one, without prompting from me, he would denounce the activities of the other. The same occurred when I visited the other. As it happened they both lived about four hours walk from Bambamarca which meant I could not be a frequent visitor. However, because of the clear views across the countryside one could always see if I was visiting the other, so I had to make a point of visiting both within a short space of time in order to maintain cordial relations.

These leaders probably thought that my contacts with the official at the Ministry of Agriculture were closer than they actually were. Their claims about each other were, therefore, what they wanted conveyed back to him. There were some quite serious problems in these sectors with a group of peasants who did not accept the organization of the *grupo campesino,* the payment of the agrarian debt or a cooperative plan for a eucalyptus plantation. The first time I visited one of these conflict-ridden sectors, two hundred eucalyptus saplings had been burnt down. On the same occasion, in the course of the journey I came across some men constructing a bridge across a stream. One of them took me for an official from the local Sub-Prefect's Office and denounced the members of the neighboring *grupo campesino* for failing to help with the building of the bridge. There were constant problems of peasants damaging their neighbors' crops, invading land, and squabbles over rights to firewood. While the two leaders appeared to be attempting to deal with these problems through the *grupos campesinos,* the local Civil Guard and the *Liga Agraria* (Agrarian League),[5] it was clear that in many instances they were causing the problems themselves.

I concentrated my fieldwork in the sector of the former estate nearest to the village. Here, though there were plenty of conflicts over the land, relations were not so fraught. The official leadership of the *grupo campesino* was not so powerful and, though one leader was suspected of selling contraband wood and another had a large plot of land, the presence of the families who had been evicted in the 1950s had checked the growth of personal political power based on the sale of land. Their own experience of struggle and certain familiarity with legal matters had enabled them to challenge one leader when he had tried to charge for access to the land. It was in this sector that I developed the widest range of social relationships, and eventually worked as a volunteer in adult literacy classes.

RESEARCH FINDINGS

When I came to examine exactly what I learned through participant observation, I find I can divide the data into two distinct types. On the one hand, these are the activities, social interactions and relationships that I actually observed and to a lesser extent participated in. On the other hand, there was what people told me about their relationships, their beliefs and their perceptions of the past. I, therefore, obtained data that can be labelled objective, insofar as any observer is able to collect objective data, and data that was subjective and attitudinal. However, perhaps more important in terms of understanding attitudes and ideas were the links between ideas and action that I was able to make, through inference and association, based on the experience of living from day to day in the community.

What is most striking, when looking back on the past, is the extent to which I selected topics that were relevant to a greater or lesser degree. In my field notes I have ample documentation of conflicts over access to land that arose in the wake of the land reform and of the rivalry between leaders. Also present, but not taking such a prominent position in my thesis, are the general features of socal, political, ritual, and economic life. These include observations on a range of subjects, from witchcraft beliefs, religious ritual, common-law marriage, artisan production, alcoholism, violence in the family, the work of Catholic lay preachers, the growth of Protestant sects, attitudes to death—in effect, all the detail of social life which added together makes participant observation a holistic experience. On the other hand, topics that would have stood as a research project in their own right go virtually undocumented. Among these I would include the expansion of the state apparatus in the rural areas, the theology of liberation and practice in the Catholic Church, and double standards in morality. I would attribute this to my preconceived notions concerning what research themes I was examining, and the inevitable need to select from a complex social reality.

Political and Ethical Problems

Political and ethical problems arise with many kinds of social science research, regardless of the research method used. The intrusion of the researcher into the private life of the individual raises problems of confidentiality. Furthermore, researchers have to consider not only the intended but also the unintended consequences of the dissemination of their research findings.

It may well be argued that with a formal interview technique the rules of the research game are clearly understood. The structure of the interview delineates what information is classified as data and collected. The interviewee can refuse to answer, challenge the assumptions of a question or refuse to be interviewed altogether. The fact that data is collected from a number of informants, through a relatively formal method means the information is more impersonal than the highly personal materials collected through participant observation. Not only is the relationship between the researcher and the subject more intense with participant observation, but the objects of the participant observer's interest are rarely free from observation, and may not even realize that they are being observed. Therefore, the greater intimacy established in the research relationship requires greater responsibility in the disclosure of information.

While interviewing techniques are appropriate in some circumstances in the literate culture, they are not appropriate where people are illiterate and the interviewer and interviewee are of very different social status. However, researching peasants, proletarians and shantytown dwellers in under-developed countries raises much more serious problems from the ethics of research work, which are not concerned simply with the question of research methodology but with the choice of subject for investigation. This is particularly the case where political guarantees do not exist or may be withdrawn by a change of government. Under these circumstances, is any research on the dominated classes appropriate? George (1976) argues that there is no justification for studying the poor and oppressed. She writes:

> *Study the rich and the powerful, not the poor and powerless.* Any good work done on peasants' organisations, small farmer resistance to oppression, or workers in agribusiness can invariably be used against them. One of France's best anthropologists found his work on Indochina being avidly read by the Green Berets. The situation becomes morally and politically worse when researchers have the confidence of their subjects. The latter then tell them things the outside world should not learn, but eventually does. Don't aid and abet this kind of research . . . Let the poor study themselves. They already know what is wrong with their lives and if you truly want to help them, the best you can do is to give them a clearer idea of how their oppressors are working now and can be expected to work in the future (1976, p. 289).

While this may be an overstatement of the potentially detrimental uses of research data, it underlines the class and neo-colonial dimensions of fieldwork conducted by academics from the industrialized countries on the oppressed of the underdeveloped countries. What to the researcher may simply be an academic exercise in collecting data leading to a higher degree and hopefully employment may, in the context of relations between classes and between countries, be an extremely useful process of intelligence gathering. It is not that the anthropologists and others conducting research on the Third World deliberately set out to collect data that might be useful for military and political purposes, but rather that they are insufficiently aware of the consequences of the data that they gather in the context of a socio-political system that is unlike the one with which they are familiar. What is merely ethnography at one point in time may have very different significance at another.

Had I worked in what became the emergency zone of Ayacucho, where the Peruvian army was conducting counter-insurgency operations against the guerilla group *Sendero Luminoso* (Shining Path),[6] I would be seriously concerned about the consequence of the dissemination of any of my data. However, circumstances change over time. James Petras, reviewing a Peruvian military report on *Sendero Luminoso* argues that evidence shows the organization is growing in politico-military capacity, expanding its bases of operations and developing internal and international support to the extent that its influence in urban areas and the northern and southern highlands has increased (*The Guardian,* Nov. 11, 1986). In this context, the real dangers of my field data lie not in the information about the politics of the 1960s nor the dealings in land of the 1970s, for what I have been told about the former is known by the security forces, and the latter was known and condoned by the authorities. On the contrary, it is the trivia of everyday life, material and attitudinal, so assiduously collected by the participant observer and the powerful insights this gives on social processes which are potentially most damaging.

SOME CONCLUSIONS

In the preceding pages I have outlined some of the reasons why participant observation, combined with an examination of documentary sources, was the only research method that could be used in a fieldwork situation characterized by conflict and a recent history of political struggle and some of the problems of interpretation which arise when sources of data are contradictory. I have indicated some of the advantages of using a holistic method in a situation in which a survey method was impossible. One of the great strengths of participant observation is that living in a community allows the researcher not just to observe but to understand through practical experience the social problems and processes being studied. Nevertheless despite the levels of the researcher's involvement in the life of the community, aspects of its history will not be revealed, or will not be articulated in terms that the observer understands.

The benefits of participant observation are that, compared to more formal research methods, the researcher gains a "feel" for a community and an understanding of social processes that is mediated by experience. My fieldwork gave me an invaluable understanding of the theoretical issues involved in studying the peasantry and helped greatly in interpreting other sources of data, such as the Land Judiciary documents. I was allowed a privileged insight into the life of a community whose history and dignity in the midst of extreme poverty made a deep impression on me. But the strengths of participant observation as a method in no way detract from the serious reservations I have concerning the use of data derived from it. These reservations arise not so much from the research method itself, but from using it to study members of an oppressed class. Although participant observation gives powerful insights into the life of a community, and in particular to the consciousness and social perceptions of the members of that community, the researcher carries a considerable responsibility toward the people studied. If that conflicts with the need to produce academic "goods," than the first responsibility must be toward the people studied. What is possible is a greater understanding by researchers of the political implications and potential uses of their findings and this requires a major refocusing of priorities in the definition of research problems.

ACKNOWLEDGMENT

I am grateful to Jenny Hockey for comments on an earlier draft of this chapter.

NOTES

1. See Blanco (1972, pp. 56-61) for a discussion of how the Trotskyist concept of dual power was applied to rural struggles. He argues that power in a rural area is clearly concentrated in the landlord because he is not just owner of the means of production but also appoints local officials and enforces law and order. Therefore, when peasants successfully organize to fight for better working conditions, they displace the rule of the landlord in other respects. A situation of incipient dual power at the local level is an unstable one, since ultimately only one class can exert decisive power.
2. FEDECC (1975) gives an amusing account of this incident.
3. "For if the co-operative is not successful here, it will be impossible to implement plans in other haciendas, because of the bad example set in Chala. Besides this, the peasants throughout the zone are watching developments here" (Land Judiciary File, October 30, 1970).
4. "The Union used to be a great organisation, but then later the functionaries of the state made us disband ourselves" Peasant account (FEDECC 1975).
5. The Ligas Agrarias were set up by Sistema Nacional de Apoyo a La Movilización Social (SINAMOS) in the rural areas in opposition to existing class-based peasant organizations such as the Confederación Campesina del Perú to mobilize support for the military régime. Though using the ideology of participation, they were in fact organizations of co-option.
6. Sendero Luminoso is a Maoist guerilla group with roots in the peasant communities of Ayacucho and began armed actions in 1980. By December 1982, the ineffectiveness of the police in containing Sendero resulted in President Belaúnde creating a Military Emergency Zone centered on Ayacucho. Since 1980 some 5,000 people have died in the region. 2,000 have "disappeared" and there are 1,000 political prisoners (Harding 1984).

REFERENCES

Alegría, Ciro
1973 *El Mundo es Ancho y Ajeno.* Lima: Varona.
Alegría, Claribel
1987 *They Won't Take Me Alive.* London: The Woman's Press.
Blanco, H.
1972 *Land or Death. The Peasant Struggle in Peru.* New York: Pathfinder Press.

Burton, F.
1978 *The Politics of Legitimacy. Struggles in a Belfast Community.* London: Routledge and Kegan Paul.

Cabezas, O.
1987 *Fire from the Mountain. The Makng of a Sandinista.* London: Jonathan Cape.

FEDECC
1975 *Luchas Campesinas en Cajamarca. Los Pobres del Campo Hacemos Nuestra Historia.* Cajamarca: Federación Campesina de Cajamarca.

Ferner, A.
1983 "The Industrialists and the Peruvian Development Model". In D. Booth and B. Sorj (eds.), *Military Reformism and Social Classes: The Peruvian Experience 1968-80.* London: Macmillan.

George, S.
1976 *How the Other Half Dies. The Real Reason for World Hunger.* Harmondsworth: Penguin.

Godelier, H.
1977 *Perspectives in Marxist Anthropology.* Cambridge: Cambridge University Press.

Guevara, E.C.
1968a *Bolivian Diary.* London: Jonathan Cape/Lorrimar.
1986b *Reminiscences of the Cuban Revolutionary Struggle.* London: Monthly Review Press/George Allen and Unwin.

Harding, C.
1984 *Origins and Development of Sendero Luminoso.* Paper presented to the Peru Seminar held at the University of Liverpool, December 17-18.

Hobsbawm, E.J.
1971 *Peru: The Peculiar Revolution.* New York Review of Books, Vol. 16.

Kautsky, K.
1972 *La Cuestión Agraria.* Lima: Universidad Nacional Mayor de San Marcos. (First published in German in 1899 as *Die Agrarfrage.*)

Lenin, V.I.
1972 *Collected Works, Volume III. The Development of Capitalism in Russia.* Moscow: Progress Publishers.

Quijano, A.
1967 "Contemporary Peasant Movements." In S.M. Lipset and A. Solari (eds.), *Elites in Latin America.* New York: Oxford University Press.

Samaniego, C.
1978 "Peasant Movements at the Turn of the Century and the Rise of the Independent Farmer." In N. Long and B. Roberts (eds.), *Peasant Cooperation and Capitalist Expansion in Central Peru.* Austin: Institute of Latin American Studies, University of Texas Press.

Sulmont, D.
1975 *El Movimiento Obrero en el Perú: 1900-1956.* Lima: Pontificia Universidad Católica del Perú.

NOT WAVING, BUT BIDDING: REFLECTIONS ON RESEARCH IN A RURAL SETTING

Kristine Mason

This chapter contains a reflection upon some of the methodological and theoretical concerns that I addressed while conducting research into gender and schooling within an English rural setting. Consideration is given to the origin of the research and its conduct, and to the issue of the generation and testing of theory.

The research project derived from my long-standing interest in the substantive issue of women's education. I also had a particular interest in qualitative methodology that stemmed from my earlier studies in social anthropology. In 1979 I successfully submitted a research proposal to the Social Science Research Council (S.S.R.C.) for a Linked Studentship which was to be attached to the University of Aston in Birmingham's research project that focused on the social effects of primary school re-organization in rural areas of England. Being linked to a rigorous and committed research team was both

Studies in Qualitative Methodology, Volume 2, pages 99-117.

ISBN: 1-55938-023-3

helpful and stimulating, and I did not suffer from the feelings of isolation and despair that so often afflict novitiate Ph.D. students. These feelings were to come much later! By this later time the S.S.R.C. no longer existed, neither did Aston's Department of Educational Enquiry—both victims of "reorganization," and my joint supervisors, Margaret Small and Bob Meyenn had left the university for, respectively, Merseyside and Wagga Wagga. Fortunately, I was reallocated to Geoffrey Walford and Henry Miller, and my last tutorials as a very part-time student at Aston were conducted among the pale grey carpeted walls of the brand new Management Center.

The original aim of my research proposal was a broad one—that of exploring the context of girls' schooling in a rural setting. This was to include conducting an ethnographic study in a rural location during the spring and summer of 1980. It was decided that the research should not be entirely school based and so my research aims were somewhat removed from what Delamont and Atkinson (1980, p. 148) then referred to as "the recurrent preoccupation of the British sociologists (which) has been the organization and negotiation of everyday life in schools and classrooms."

They went on to argue that:

> The sensitivity of British sociologists to the negotiations of everyday life within schools and classrooms has tended to obscure relationships between schooling and local culture, local structure and so on (rather than the very broadest categories of social structure) (p. 148).

Like Delamont and Atkinson I considered that the local context of the schooling of rural school students was an important area of study. The word "context" like that of "setting" is vague, but I considered this to be advantageously permitting rather than detrimentally restricting. I regarded both local social and economic structures, and their relationship to actors' perspectives to be part of the formation of the context of pupils' schooling in the area selected for study.

Access to somewhere to live in the selected rural area under study was an important early consideration. As I stated earlier, my project was "linked" to an ongoing research project at Aston; as part of the Aston research, Bob Meyenn had been conducting some interviews and making observations in Headleigh[1] village primary school, the village in which, following discussions with members of the research team, I had hoped to conduct the research. Headleigh seemed a particularly suitable location for the ethnography, especially because it contained both a primary and a secondary school. The village, with its population of some five hundred "locals" and "incomers" was situated in a lovely valley. Most of the land was down to pasture, and about a quarer of it was devoted to the production of arable crops—wheat, barley, and oats. Unlike the large agribusiness, capitalist farming concerns that dominate many other rural counties, around Headleigh it was the much smaller "family farm" that predominated.

Tourists in the valley would probably not have chosen to stop at Headleigh: its development of small new modern houses which flank one side of the village (referred to as "the pigeon houses" by local farmers), and a small estate of council housing, do not consistute elements of the rural idyll. However, the valley in which the village was situated is very beautiful, and a brief visitor to the area might very easily subscribe to the ideological view of pastoral England so well described by Saunders et al. (1978, p. 63) as being "the repository of all that is stable, in memorial, harmonious, pleasant and reassuring in modern society."

While working in Headleigh primary school, Bob Meyenn had related well with one of the teachers, Carol Smith, and had asked her whether she and her husband Eddy Smith could accommodate me in their house on a small holding of five acres just outside the village. She was seemingly very agreeable to this suggestion, and I went to visit her myself. It was only then that I realized Bob Meyenn had "forgotten" to mention that I would be accompanied by my 3-year-old daughter. After the initial surprise at this omission, Carol was willing to take us on, and it was through her that I made childminding arrangements and learned that I could use the village playgroup that operated in the village community center for three mornings a week.

Through my relationship with the Aston research project I had ready access to both the primary and secondary schools in the village. As an initial clarification of my research area I took copies of its general aims with me to the schools to discuss with the head teachers. Through my access to the village playgroup I began to make contact with other women, and eventually I attended meetings of the Women's Institution, the Parish Council, and the Youth Club. Having gained initial access to various sites it soon became clear that in order to work fruitfully, I would need to continuously negotiate access, and, because my fieldwork extended outside the boundaries of formal organizations, I felt that the process of continuous negotiation of access was heightened.

Before extending my fieldwork beyond the schools, I endeavored to develop some form of sampling procedure for studying "people and events." As I was keen to interview parents, I asked the heads of the two schools whether they could let me have names of parents in the cateogries of "local" and "newcomer" (terms I discovered that were used by residents in the area) so that I could make arrangements to meet and interview members of the two "categories." After some reflection, however, I became aware that such an approach in the area would have been extremely unsuitable. One woman had told me how once a researcher had knocked on her door with a questionnaire and how she had "answered all the questions as quick as possible just to get rid of him. I didn't stop to think what I was saying." Like Lynda Measor (1985, p. 57), I considered that "the quality of the data is dependent on the quality of the relationship you build with the people being interviewed."

For this reason I decided to "follow leads" in terms of the contacts that I built up during my stay in Headleigh. In this respect my research procedure somewhat resembled that of Hoffman:

> Abandoning her original research design—based on interviewing a representative example (sic) from different institutions—Hoffman therefore started to select informants on the basis of social ties (Hammersley and Atkinson 1983, p. 61).

Hoffman's degree of abandonment would not have been appropriate for me, since to have only followed "social ties" would, in the location where I was researching, have meant that either "locals" or "incomers" would have been relatively omitted, for as the headteacher of the primary school told me, "there is a chain-link fence surrounding the indigenous community." What I did, therefore, was to develop relationships through my involvement in the playgroup, the youth club and such organizations as the Women's Institution, all of which were attended by locals and incomers.

During the course of the ethnography, my interviews, conversations and observations led me to focus my attention upon the "social" lives of young people in the area and what their employment intentions were, since it began to appear that these factors were closely related to their perspectives on schooling. This reflection led me to focus progressively upon careers advice and employment, and further education opportunities in the local market town of Beckford, some twelve miles away from Headleigh. This moving of my research sampling to another area found support in the view expressed by Hammersley and Atkinson (1983, pp. 43-44).

> Whilst it may seem innocent enough, the naturalistic conception of studying fields and settings discourages the systematic and explicit selection of aspects of a setting for study as well as movement outside it to follow up promising theoretical leads. It is important to remember that the process of identifying and defining the case under study proceeds side by side with the refinement of the research problem and the development of the theory.

As well as sampling particular groups of people—pupils, teachers, and parents—much time was spent in participating in as wide a sample of events and activities as possible.

In Headleigh village I observed meetings of the Parish Council, the Community Center Committee, and the Women's Institute. At such meetings I could see who occupied positions of status within the village and I could also observe the manifest tensions between "locals" and "incomers." At mixed sex meetings the sexual division of labor was clearly evident. I was an active participant observer in the playgroup and in the youth club, and, in addition, I observed a wide range of activities including a school play, a barn dance, a young farmers' rally, a school sports day, and a fashion show in the community center hall.

My main informants were Carol and Eddy Smith in whose beautiful old house I lived with my daughter. Carol taught in the primary school and Eddy worked full time on the 5-acre holding caring for small numbers of cows, sheep, and goats. They had lived in the area for five years and were thus defined by themselves and others as "incomers."

Another main informant was Derwent Hall, a "local" who farmed 18 acres, dairy and sheep, and cut wood for a living. Although he was disparaging about incomers in general he told me he had been very impressed by Carol and Eddy Smith's commitment to hard work on their farm. Derwent Hall lived a mile and a half away, and he was a very regular evening visitor to Carol and Eddy's home when he would come into the kitchen wearing overalls, cap and Wellington boots and lean against the wood burning oven drinking pints of sweet tea. Clearly there were gaps in the "chain link fence."

For me, self-presentation was a very important factor in gaining and maintaining access to all of these sites. Like Measor (1985), I took special care to dress appropriately. I did not wear the longer skirts and large jewelry I favored so that I would not look like "one of those hippies" which was how locals disparagingly referred to the incomers, aged mainly in their thirties, who had moved into the area to seek various degrees of "self sufficiency."

When I was asked what my research was about, I would make a reply along the lines that most educational studies have been conducted in urban areas and that I was interested in education in rural areas. In most instances, however, I found that,

> Whether or not people have knowledge of social research, they are often more concerned with what kind of person the researcher is than with the research itself (Hammersley and Atkinson 1983, p. 78).

A similar point has been stressed by Jahoda and Sellitz (1959) who said that a researcher's personal attributes "entirely unrelated to his scientific skills" may be decisive factors in the community's tolerance of his or her activities.

In his doctoral thesis Meyenn (1979) says of himself that being an Australian, "I was something of a curiosity." I too was undoubtedly something of a curiousity in that I was (am) a six foot woman single parent who was conducting research accompanied by a young daughter. [Though unlike Corrigan (1979) my height did not deter me from trying to use what he terms "unobtrusive participant observation."] Like Meyenn, I could argue that this being different gave me certain "stranger value" that could, of course, be put to good use by an ethnographer who may then ask questions and seek clarification about "the obvious."

Central to my relations in the field was the fact that I was a woman and a mother. Although some of the personal stress involved in combining the latter role with that of researcher was often keenly felt by me, I think on reflection that there was considerable advantage to be gained. A similar view has been expressed by Bennett and Kohl (1981, p. 94) who wrote of Kohl,

> Her own status as wife and mother was an important element in the exchange of information about children and aspirations for children.

I certainly felt that as a participant observer in the playgroup, I had very ready access to other mothers who would willingly discuss their views of their children, schooling, and relations between men and women.

With regard to my marital status I let it be known that I had recently divorced and that my former husband worked abroad. Although, as stated earlier, a single parent researcher must have seemed unusual, I felt enormously encouraged when the blacksmith in Headleigh told me over tea at his home that, in terms of carrying out my work in the village, "I think your best qualification is being a mum." He went on to criticize an earlier survey that had been conducted in which the researchers had evidently been, in his words, "seen as students trying to get their degrees. There was a bit of a joke, 'have you been done yet?' and people saying they'd pretend to be out when they called."

With regard to the question of gender, I would agree with Morgan (1981) that one's gender has too often been viewed as a "source of difficulty." He says,

> gender differences in fieldwork are not simply a source of difficulties such as exclusion from important rituals, or in my case, exclusion from all important interactions in the toilets, but are also a source of knowledge about the particular field. The "participant observer," in short, was a gender identity (p. 91).

I was very aware of my gender identity as a source of knowledge. For example, I noted how my bottom was pinched by an old local farmer in the pub following a meeting of the Community Center Committee. On another occasion my notes refer to a young a farmers' rally that I observed with my supervisor Margaret Small,

> We left the main area of the rally and wandered over to look at some of the agricultural machinery. There was a group of men examining it and as we approached they stopped their talking and looked coldly at us. We felt frozen out. Clearly we had trespassed on male territory.

My status as a qualified teacher appeared to enhance my field relations with heads and teachers. One occasion in the primary school is referred to thus in my notes when a teacher rushed into the staffroom and said to me, "'Quick, can you take over my class for a bit? Tell them a story or something.' I suddenly found myself in front of thirty children."

When I asked the secondary school science teacher (who I'd learned was the village youth club leader) whether I could attend the club regularly, he responded extremely favorably when I said that I had once been employed as a youth worker. I thus became a participant observer in the youth club and through this route I was easily able to attend youth club activities and discos. I continued to build up a number of contacts in a perhaps rather opportunist way, as the following diary note suggests, following a "coffee morning" with two women in the village.

> They mentioned several names I'll take them up on—for more contacts.

I found that great flexibility was demanded in my use of time. I had read about the problems for the ethnographer in planning the use of time in advance, (for example, Jahoda and Sellitz 1959), yet on occasions I felt extremely frustrated, as indicated in the following diary entry:

> One day up the next down. Yesterday I met three farming families, today the one I'd arranged to meet said would I get in touch later.

Sometimes, in order to maintain positive field relations, I had to spend time in an activity that I would not necessarily have chosen myself. For example, the wife of a former headmaster of one of the schools insisted that I accompany her to a meeting of the Association of Teachers of Domestic Science of which she was a member. (Although I was able to make some interesting observations at the meeting.) On other occasions my efforts to maintain good field relations involved me in such activities as eating an enormous plate of liver casserole that had been prepared for my arrival to an interviewee's home. I do not like liver and I had eaten my evening meal some twenty minutes previously.

Assessing how well access and field relations have been established and maintained is difficult, and can probably best be evaluated in terms of the data collected. I did feel, however, that by the end of my stay in Headleigh that when I had been invited to "incomers" parties where there were no "locals" and to a local farming family's wedding party where I observed no "hippies" or "retired townies," that I had achieved some measure of success. Having said this, however, I considered at the time, and even more so after greater reflection, that the ethical issues related to establishing and maintaining "good" field relations, are enormous.

One particular issue that concerned me was that once good rapport and confidence has been established between the researcher and "the researched," the latter may well reveal much more than they might originally have intended,

(Janes 1961). Were my efforts to establish good rapport and friendly relations merely a means of catching people off their guard so they would feel free to disclose more and more?

For many years, writers on qualitative methodology have made reference to the question of ethics in the conduct and reporting of participant observation (e.g., Denzin 1970, Erikson 1970). Certainly this is an area that has occupied the minds of many field researchers. Anthropologists such as Beattie (1960), Powdermaker (1967), and Turnbull (1972) have presented vivid accounts of the ethical issues that confronted them during their fieldwork. They had to make decisions about what they should do with the knowledge they had acquired about illegal practices, such as the distilling of alcohol, and about how they should behave in the presence of male physical violence toward women. Turnbull describes his efforts to provide for the dying lk.

Despite Frankenberg's often quoted dictum that researchers should only be carrying out their work with the full knowledge of those being "researched," some sociologists have employed covert ethnographic techniques in order to pursue their research, believing there to be no other means of studying illegal practices (e.g., Humphreys 1970).

Within the sociology of education, researchers such as Hargreaves (1967), Willis (1977), and Walford (1986) have made reference to their knowledge of illegal practices occurring either within the school setting or conducted by the pupils elsewhere. Hargreaves (1967) has referred to his being offered stolen property and how boys at Lumley would relate their latest stories of theft to him. By contrast, Walford (1986) gives almost nothing away. A number of researchers (e.g., Burgess 1983), have made reference to how they found themselves in awkward ethical situations when it came to "permitting" officially illegal practices to occur in their presence.

From my experience, although I was not faced with any of the harsh dilemmas encountered by Beattie, Powdermaker, and Turnbull, and although I would not support the covert methods employed by Humphreys, I was certainly aware of constantly facing ethical questions. Just as access to, and within, the field is permanently negotiated, so too I would argue, are questions of ethics—both during the fieldwork itself and in the subsequent writing up and dissemination of the research.

During the fieldwork I became aware of details of covert practices such as "moonlighting," "tax-dodging," and various details of "gossip" that I was asked to keep confidential. After one long and informative discussion with the headteacher of the secondary school I was asked by him to keep what had been said "within these four walls," despite the fact that I had obviously been taking notes throughout the discussion. In hindsight, I think I should have asked him to be specific about what should not be disclosed rather than ruefully acquiese to keeping the whole session confidential.

As stated earlier, although I would not support the idea of covert research, I was very much aware that some of my note-taking was covert. It is now commonplace for ethnographers to relate how they have dashed into the seclusion of the toilet to write up a few notes (in contrast to Humphreys who, presumably, dashed out of the toilet to write his). This was certainly one of the venues that I frequented to jot down some observations or thoughts. I was covert too, in that because my ethnographic study was not confirmed to a single institution in which I was known to be a researcher, I could not reasonably preface every new encounter in every new situation with a statement as to my reason for being there.

During the ethnography I employed a range of methods of investigation. Although I observed various activities and events, I would not say, however, that I was ever purely an "observer," for I fully support Hammersley and Atkinsons' (1983, p. 16) point that:

> all social research takes the form of participant observation: it involves participating in the social world, in whatever role, and reflecting on the products of that participation.

Over the years, several social scientists have made reference to different "types" of participant observation; for example, Schwartz and Schwartz (1955) suggested that there is a continuum from the "passive" participant observer role in which the researcher has little opportunity to share in the life of those being observed, to the "active" participant observer role in which the researcher attempts "to integrate his role with other roles in the social situation."

My own range of participation varied from the relatively passive, for example, as "onlooker" at the bidding in a livestock market, to active participation, such as being involved with other mothers in taking playgroup children swimming.

Conversation was an important means of social investigation (although as stated previously there are particular ethical problems involved with it) and I would maximize opportunities for conversation in a number of ways. For example, I would give lifts and literally go out of my way as illustrated in the following note:

> Saw Susan walking down Hill Lane. I gave her a lift into Beckford Town although I hadn't intended to go just then.

Interviewing was the main means of data collection during the ethnography. Informal interviews were conducted by arranging convenient times in advance, although, as stated earlier, these times were not always adhered to. While interviewing I would have with me a number of headings that I wanted to be sure to ask about; however, I was keen also for interviews to "take their own course" to a considerable extent, for I felt along with Ann Oakley (1981, p. 38) that:

> the paradigm of the 'proper' interview appeals to such values as objectivity, detachment, hierarchy and 'science' as an important cultural activity which takes priority over peoples more individualized concerns.

The setting undoubtedly influenced the level of formality of the interview. There was almost a contniuum of formality of setting, ranging from interviewing a woman in her kitchen while she plucked a chicken and our children played, to interviewing the head of the business studies department in his office in Beckford Technical College in the local market town.

A considerable range of documentary material was examined and collected during the course of the research. This included bus timetables (indicating the rarity of buses that was so often complained about), District Plans, letters shown to me by residents, wages slips, local newsletters (in which "Nurses Notes" detailed aspects of local health) and archival material.

The principal means of recording the ethnographic data were those of notebook and a cassette recorder. In recording data I attempted to bear in mind Frankenberg's dictum referred to earlier, that "if the observer cannot participate with the knowledge and approval of the people to be studied he should not be there at all" (Frankenberg cited by Hargreaves 1967). However, I was frequently reminded of Hargreaves' caveat that "The difficulty is that most of the people being studied could not appreciate that many of the apparently trivial things they said or the confidences they related are of social significance" (Hargreaves 1967, p. 199).

As well as the notes that I jotted down during the process of observing interviewing, or as soon as practicably afterward, I also kept diary notes in which both personal feelings as well as reflections were recorded. In the process of field research the researcher can feel very isolated. I found that writing the diary provided some support as well as the means of separating my personal feelings from the ethnographic account. An example illustrates the point from the diary notes:

> Walking back in the rain to collect Charlotte from Jenny's I felt despondent. Another day, some more notes. But before leaving the community centre I made a note of the W.I. President's name. I learned from Jenny she lived next door. I called in and am now invited to the W.I. meeting next week, and for a talk with her. Suddenly things seem better.

The interviews I conducted with primary school pupils, parents, teachers, youth leaders, and careers advisers were recorded in my own brand of shorthand after I had secured the "respondent's" agreement. The interviews with groups of fourth and fifth year secondary school pupils in the village were tape recorded. In these latter group interviews I wanted to be able to reflect and to "progressively focus" during the interview. I took notes during these interviews and immediately afterward wrote down some thoughts that had occurred to me about what had been said. During the interviews the pupils did appear to be reasonably at ease with the recorder being switched on. Sometimes they asked for it to be played back, and on one occasion a boy said the machine had stopped working some twenty minutes previously, and there was great hilarity (and on my part relief) when I discovered he'd been "joking." The process of making my jottings legible and of fully transcribing the tapes, although time consuming and laborious, was considerably leavened by the apparent willingness of the respondents to express their views freely.

The organizing of the ethnographic data was carried out by categorizing the notes and transcripts into 43 separate headings. This system of categorizing and subsequent indexing served as a very useful means of retrieving the material. Analysis of the ethnographic data was ongoing and, as will be discussed later, it was through this process that my explanatory theory developed.

As well as employing ethnographic methodology in the research process, quantitative methodology was also utilized. This took the form of administering and analyzing a questionnaire prepared for all fifth form students in the Headleigh secondary school. The questionnaire was designed to ascertain whether a number of hypotheses that had been generated during the ethnography were quantitatively verifiable.

The questionnaire was not selected purely as a method of investigation to either "confirm" or "disconfirm" the hypotheses generated from the ethnography concerning girls' and boys' schooling in the area. Such an intention would, by implication, have conferred some greater validity to the quantitative approach. Rather, it was intended to serve as a "partial testing" (Woods 1985) of locally generated theory. The questionnaire method was not, therefore, regarded as possessing some supremacy over ethnography, for to claim that one method is "better" than another is, as Trow (1957) has suggested, the equivalent of saying that the scalpel is better than foreceps. Clearly the relevant method needs to be employed for addressing the particular research problem, or "stage" of the research process.

The value of the questionnaire as a testing strategy rested mainly on the fact that the questions it was designed to address were well grounded in the ethnographic study itself, rather than such questions being generated out of some armchair positivist's preconceptions. The questionnaire method of investigation served to provide the means of ascertaining whether there was a correlation between discrete factors, and whether any gender differences related to schooling and life in the rural area were statistcally significant. My research was, therefore, rather unusual in that quantitative methods were used for the partial testing of certain locally generated hypotheses.

The prime value of the ethnographic stage of the research was clearly that of generating hypotheses in relation to the schooling of girls in a rural context, and in so doing generate sociological theory. The value of ethnography in generating hypotheses has long been recognized by researchers who have employed the method. For example, Whyte (1945) emphasized the point that "ideas should grow up in part out of our immersion in the data," and Humphreys (1970) stressed that "hypotheses should develop out of ethnographic work rather than provide restrictions and distortions from its inception."

To say that hypotheses emerge during the enquiry is not, of course, to suggest that the researcher enters the field with a completely blank mind. As anthropologists such as Beattie (1960) have long been saying, it is not possible to undertake any investigation "lacking any notion" of what is to be found out or how to set about the task. Sjoberg and Nett (1968) state that the researcher begins with a broadly defined problem or set of problems that may be radically revised.

It is important too to stress the point that hypotheses are *generated* rather than that they merely *emerge* out of the data collected. As Woods (1985, pp. 51-52) states,

> Theory does not simply 'emerge' or 'come into being' . . . . at some stage there must be a 'leap of the imagination' (Ford 1975) as the researcher conceptualises from raw field notes.

Clearly, then, the *interpretation* of the data by the researcher is critical. Nash (1984, p. 27) illustrates this point when he says "There is no way of determining what behaviour is 'resistance' . . . and what is not other than by a process of interpretation."

Interpretation is, of course, a permanent feature of everyday interaction. One woman farmer in the area where I was working told of how women farmers were so rare that she lost the chance of buying some good stock at Beckford Town market when the auctioneer assumed that she was waving to a friend, rather than making a bid at the sale. In my work I found that the process of interpretation and the revision of ideas was a permanent and demanding feature of the conduct of the ethnography itself, as well as being emphasized later in this paper. The use of questionnaires and statistical analyses was intended as a means of further testing the hypotheses that I derived during the ethnographic stage of the research. Although, while engaging in these procedures, I was not familiar with work in educational research that had attempted to generate and test hypotheses in this way, it has since become clear that certain ethnographers (e.g., Woods 1985) would consider this to be a useful approach. According to Woods, questionnaires can "act as a partial test of theories generated locally" (p. 68). With regard to theory construction and quantification, Hammersley and Atkinson (1983, p. 19) say of the development and testing of theory that,

> It is this that marks it off from journalism and literature, even though it shares much in common with these other pursuits (Strong 1982). Moreover, the idea of relationship between variables that, given certain conditions, hold across all circumstances seems essential to the very idea of theory (Willer 1967). Quantification, as an aid to precision, goes along with this too; though this is not to excuse the indiscriminate quantification that positivism has sometimes encouraged.

In this respect Hammersley and Atkinson are echoing the views expressed nearly twenty years previously by Glaser and Strauss (1967, p. 6) who said,

> Description, ethnography, fact-finding, verification (call them what you will) are all done well by professionals in other fields and by layman (sic) in various investigatory agencies. But these people cannot generate sociological theory from their work. Only sociologists are trained to want it, to look for it, and to generate it.

Glaser and Strauss also make a point concerning grounded theory that,

> The theory should provide enough categories and hypotheses so that crucial ones can be verified in present and future research; they must be clear enough to be readily operationalised in quantitiative studies where these are appropriate (p. 3).

Although ethnographers such as Woods appear to stress the "local" nature of theory generation, the statements by Glaser and Strauss, and Hammersley and Atkinson appear to give credence to the broader testing of theory.

It should be made clear that while the original design of my research was based upon conducting ethnographic work, exigencies of my personal life (moving and finding full-time employment) had an affect on the design, so that a longer period than was originally anticipated was necessary for the analysis of field notes, and the design, piloting, administration and analysis of the questionnaires. I would also make the point that "research design" in relation to ethnography implies a notion of process rather than merely an original formulation of the research procedure. As Agar (1980) says of the process of research in ethnography,

> The process is dialectic, not linear, such a simple statement, so important in capturing a key aspect of doing ethnography (p. 9).

The review of the literature on schooling in rural areas that I conducted prior to the ethnographic fieldwork had revealed that such studies had employed an implicit functionalist theoretical perspective. This perspective stood in marked contrast to the theoretical perspectives that had been developing and informing "mainstream" sociology of education during the seventies (e.g., Young 1971; Young and Whitty 1977; Bowles and Gintis 1976; Karabel and Halsey 1977; Bourdieu and Passeron 1977).

Similarly, studies in schooling in rural areas had either ignored or completely undertheorized issues of gender. Again this was in marked contrast to the development of feminist critiques of schooling within the sociology of education during the seventies (e.g., Kuhn and Wolpe 1978; Deem 1978).

One of the main areas of development in feminist theory during the seventies had been the analysis of the concept of patriarchy. Delphy (1984, p. 139) has stated:

> The introduction of the term 'patriarchy' and its widespread use . . . are due to the feminist movement of the 1970s . . . . Before the new feminism the term 'patriarchy' had no explicit meaning, and above all no explicit political meaning. This is not surprising. It is the nature of patriarchy—as of all systems of oppression—to deny that they are such.

The above brief reference to paradigm shifts within the sociology of education during the seventies, and to the development of feminist theory during the same period, are made in order to indicate how I had been "theoretically informed" prior to embarking upon my research. As Roberts (1981, p. 17) has stated,

> a step which is frequently omitted from description of the research process is that of providing a background to the framework within which a piece of research is conceived and developed. Providing such a framework makes explicit the paradigms within which the research is set.

Having made this point, however, one can still see the strength in Hammersley and Woods' (1977) argument that "Too much theory at the start can prejudice the outcome of ethnography" and that "Theory is the chief *product* of the sociological enterprise" (my emphasis) (p. 51).

One may argue that the view held by Madge (1953) some forty years ago is still relevant today, and that the following statement usefully combines the apparently rather conflicting views held by sociologists concerning the role of theory in the research process. According to Madge, the participant observer should aim for some flexible midway point between "the research worker with no frame of reference who sees much but identifies little" and "the research worker with too rigid a frame of reference (who) sees things that confirm his preconceptions" (p. 124). Madge also makes reference to the point made over a hundred years ago by Spencer (1873, p. 73):

> for correct observation and correct drawing of inferences, there needs the calmness that is ready to recognize or to infer one truth as readily as another. But it is next to impossible thus to deal with the truths of sociology. In the search for them, each is moved by feelings, more or less strong, which make him eager to find this evidence, oblivious of what is at variance with it, reluctant to draw any conclusion but that already drawn.

One could argue that Spencer is articulating a very "modern" view here with regard to the recognition of different "truths," for as Rescher (1978, p. 20) has argued, the search for:

> absolutely certain, indefeasible, crystalline truths, totally beyond the possibility of invalidation . . . represents one of the great quixotic quests of modern philosophy (cited by Hammersley and Atkinson 1983, p. 17).

Clearly, too, Spencer's emphasis on the need for "calmness" is highly relevant to today's ethnographer who reflects upon observations made in order to develop a theory which is well grounded (Glaser and Strauss 1967). Through this "calmness," the creative process (Woods 1985) of theory construction may more readily occur, than in a desperate bid to make observations fit into some pre-formed and rigid theoretical conceptualization.

While ethnographers appear to be agreed on the necessity of reflexivity in the process of theory construction, as stated earlier, it is only recently that the point has been made that the theories thus elicited should be subjected to rigorous testing. Thus, for example, Hammersley, Scarth, and Webb (1985, p. 54) have pointed to the necessity for sociologists now to both endeavor to test generated theories as well as to systematically develop them. Although one may see the force of this point of view, it is arguable that there is a problem in emphasizing the difference between systematic development of theory and its subsequent testing. In my own research I attempted to develop a theory during the ethnography and then I explored it further through the use of quantitative analysis. In an "ideal research world" with infinite resources of funding and time, one would want to refine the theory on the basis of the results of subsequent analyses, returning to the ethnographic field and subsequently re-testing the theory and so on. Thus the testing of the theory becomes an integral part of the process of theory development.

Although, as stated earlier, I was "theoretically informed" prior to conducting my research, it was only during the research process itself that I developed the theory that, in the area being studied, school was an active site of contestation for the female pupils.

Far from being resistant to school it appeared that the girls hoped to *use* school to resist the patriarchal culture of the locality, and to move away and seek interesting employment elsewhere. It also became clear that the girls were thwarted in these efforts particularly by the boys in school who tended to regard schooling as rather irrelevant and also by the "careers" advice, or lack thereof, which the girls were offered. As the ethnography continued, I became more aware of the gender differentiation which pertained concerning the three areas of the perceived value of school, the perceived satisfaction derived from living in a rural area, and hopes for the future, particularly with regard to where that future might be located—either in the same rural area or elsewhere. It appeared that such gender differentiated perspectives could be related both to the local culture of patriarchal relations, as well as to the local economy that provided little in the way of employment for women.

Through observations, discussions and analysis I developed a number of hypotheses concerning gender differentiation that I intended to test by means of a questionnaire to fifth year pupils in the secondary school during 1981. Because my personal situation changed at the end of 1980 I was unable to develop and administer the questionnaire until the Spring of 1984. Thus, unintentionally, the study became somewhat longitudinal. I did not follow up the cohort that I had previously interviewed, but chose to administer the questionnaire to the 1984 fifth year secondary school pupils in order to ascertain whether the hypotheses I had developed might be statistically verifiable. This was rather a nerve-wracking experience, as I imagined that the chances of having my hypotheses confirmed might be more remote owing to the passage of time. (Spencer's requirement for "calmness" was not always a simple one to abide by.) Had all of the hypotheses been disconfirmed, this might indeed have been "interesting" but might also have tended to suggest that my earlier fieldwork was either wanting or that I had only succeeded in producing a "snapshot" of a particular moment or, indeed, that the quantitative testing was in error. In discovering that the quantitative analyses of my hypotheses tended to confirm them, I experienced the sense of relief that the Oracle team might have experienced when, as Galton and Delamont (1985) say, they found that their two observational methods—systematic and ethnographic—"produced very similar findings" (p. 174).

The day before I left Aston to conduct my ethnography in Headleigh, a statistician from another department of the university cheerily said to me, "Good luck in the writing of your fairy tale." Perhaps it was slightly in response to this remark that I subsequently felt some satisfaction as well as relief when my ethnographically generated hypotheses on gender differentiation were shown to be statistically significant.

I have, in the latter part of this paper, promoted the idea of using quantitative analysis for the partial testing of locally generated theory (Woods 1985), and I have indicated how, in my own research, I have endeavored to to just this.

Having made this point, it is clear that the two methods are not interchangeable. The researcher needs to select the method for its suitability. Clearly, no amount of statistical analysis could indicate the great significance that the ethnographer would attach, for example, to women's bidding being taken for waving in a particular setting. Clearly, too, the ethnographer must utilize the methods that will assist in the generation of theory, for, as stated earlier, it is the development and testing of social theory which is the particular domain and challenge of the sociologist.

NOTE

1. Headleigh, like all the other names in this article, is pseudonymous.

REFERENCES

Agar, M.H.
1980 *The Professional Stranger: An Introduction to Ethnography.* New York: Academic Press.

Beattie, J.
1960 *Bunyoro, an African Kingdom.* New York: Holt, Rinehart and Winston.

Bennett, J.W. and Kohl, S.
1981 "Longitudinal Research in North America: The Saskatchewan Cultural Ecology Research Program, 1960-1973." In D.A. Messerschmidt, *Anthropologists at Home in North America. Methods and Issues in the Study of One's Own Society.* Cambridge: Cambridge University Press.

Bourdieu, P. and Passeron, J.C.
1977 *Reproduction in Education, Society and Culture.* London: Sage Publications.

Bowles, S. and Gintis, H.
1976 *Schooling in Capitalist America: Educational Reform and Contradictions of Economic Life.* London: Routledge and Kegan Paul.

Burgess, R.G.
1983 *Experiencing Comprehensive Education: A Study of Bishop McGregor School.* London: Methuen.

Corrigan, P.
1979 *Schooling the Smash Street Kids.* London: Macmillan Press.

Deem, R.
1978 *Women and Schooling.* London: Routledge and Kegan Paul.

Delamont, S. and Atkinson, P.
1980 "The Two Traditions in Educational Ethnography: Sociology and Anthropology Compared." *British Journal of Sociology and Education* (2).

Delphy, C.
1984 *Close to Home: A Materialistic Analysis of Women's Oppression.* London: Hutchinson.

Denzin, N.K.
1970 *The Research Act: A Theoretical Introduction to Sociological Method.* Chicago: Aldine Atherton.

Erikson, K.T.
1970 "A Comment on Disguised Observation in Sociology." In W.J. Filstead (ed.), *Qualitative Methodology: First Hand Involvement in the Social World.* Chicago: Markham.

Ford, J.
1975 *Paradigms and Fairy Tales.* London: Routledge and Kegan Paul.

Galton, M. and Delamont, S.
1985 "Speaking with Forked Tongue? Two styles of Observation in the ORACLE Project." In R.G. Burgess (ed.), *Field Methods in the Study of Education.* East Sussex: The Falmer Press.

Glaser, B.G. and Strauss, A.L.
1967 *The Discovery of Grounded Theory Strategies for Research.* London: Weidenfeld and Nicholson.

Hammersley, M. and Atkinson, P.
1983 *Ethnography, Principles and Practice.* London and New York: Tavistock Publications.

Hammersley, M. and Woods. P. (eds.)
1977 *The Ethnography of the School.* Educational Studies: A Second Level Course. E202 Schooling and Society Units 7-8 Block II. The Process of Schooling. Milton Keynes: The Open University Press.

Hammersley, M., Scarth, J., and Webb, S.
1985 "Developing and Testing Theory: The Case of Research on Pupil Learning and Examinations." In R.G. Burgess (ed.), *Issues in Educational Research: Qualitative Methods.* East Sussex: The Falmer Press.

Hargreaves, D.H.
1967 *Social Relations in a Secondary School.* London: Routledge.

Humphreys, L.
1970 *Tearoom Trade: A Study of Homosexual Encounters in Public Places.* London: Duckworth.

Jahoda, M. and Sellitz, C.
1959 *Research Methods in Social Relations.* New York: Holt.

Janes, R.W.
1961 "A Note on Phases of Community Role of the Participant Observer." *American Sociological Review* 26:446-450.

Karabel, J. and Halsey, A.H.
1977 *Power and Ideology in Education.* New York and Oxford: Oxford University Press.

Kuhn, A. and Wolpe, A-M. (eds.)
1978 *Feminism and Materialism, Women and Modes of Production.* London: Routledge and Kegan Paul.

Madge, J.
1953 *The Tools of Social Science.* London: Longman.

Measor, L.
1985 "Interviewing: A Strategy in Quantitative Research." In R.G. Burgess (ed.), *Strategies of Educational Research.* East Sussex: The Falmer Press.

Meyenn, R.J.
1979 *Peer Networks and School Performance.* Unpublished PhD thesis, The University of Aston in Birmingham.

Morgan, D.
1981 "Men, Masculinity and the Process of Sociological Enquiry." In H. Roberts (ed.), *Doing Feminist Research.* London: Routledge and Kegan Paul.

Nash, R.
1984 "Two Critiques of the Marxist Sociology of Education." *British Journal of the Sociology of Education* 5(1).

Oakley, A.
1981 "Interviewing Women: A Contradiction in Terms." In H. Roberts (ed.), *Doing Feminist Research.* London: Routledge and Kegan Paul.

Powdermaker, H.
1967 *Stranger and Friend: The Way of an Anthropologist.* New York: Norton.

Rescher, N.
1978 *Peirce's Philosophy of Science: Critical Studies in his Theory of Induction and Scientific Method.* South Bend, IN: University of Notre Dame Press.

Roberts, H.
1981 "Women and Their Doctors: Power and Powerlessness in the Research Process." In H. Roberts (ed.), *Doing Feminist Research.* London: Routledge and Kegan Paul.

Saunders, P., Newby, H., Bell, C., and Rose, D.
1978 "Rural Community and Rural Community Power." In H. Newby (ed.), *International Perspectives in Rural Sociology.* New York: John Wiley.

Sjoberg, G. and Nett, R.
1968 *A Methodology for Social Research.* New York and London: Harper and Row.

Schwartz, M.S. and Schwartz, C.G.
1955 "Problems in Participant Observation." *American Journal of Sociology* 60(4):343-353.

Spencer, H.
1873 *The Study of Sociology*. London: Williams and Norgate.
Strong, P.
1982 "The Rivals: An Essay on the Sociological Trades." In R. Dingwall and P. Lewis (eds.), *The Sociology of the Professions: Medicine, Law and Others*. London: Macmillan.
Trow, M.A.
1957 "Comment on Participant Observation and Interviewing: A Comparison." *Human Organization* 16(3):33-35
Turnbull, C.M.
1972 *Mountain People*. New York: Simon and Schuster.
Walford, G.
1986 *Life in Public Schools*. London: Methuen.
Willer, D.
1967 *Scientific Sociology*. Englewood Cliffs, NJ: Prentice-Hall.
Willis, P.
1977 *Learning to Labour*. Farnborough: Saxon House.
Whyte, W.F.
1945 *Street Corner Society*. Chicago: University of Chicago Press.
Woods, P.
1985 "Ethnography and Theory Construction in Educational Research." In R.G. Burgess (ed.), *Field Methods in the Study of Education*. East Sussex: Falmer Press.
Young, M.F.D.
1971 *Knowledge and Control: New Directions for the Sociology of Education*. London: Collier-Macmillan.
Young, M.F.D. and Whitty, G.
1977 *Society, State and Schooling*. East Sussex: Falmer Press.

RESEARCHING AND THE RELEVANCE OF GENDER

Joan Chandler

What is the point of writing a narrative account of research experience? One answer is that it contributes to a reflexive methodology and to a wider understanding of how research articulates with research practice. Another is that, for the researcher, it is an exercise in self-criticism and an opportunity for self-justification. The confessional elements tie into the egocentricity of the research enterprise and into the intellectual insecurity of being a Ph.D. student. Although the exposition of tell-it-all accounts is now a popular methodological exercise (Bell and Newby 1977; Burgess 1984), there are limits to self-revelatory writing. There is danger in assuming that the style of the narrative account makes for greater honesty in description for it also provides avenues for alternative humbug and misrepresentation.

I have tried to avoid giving testimony to how, though initially unsure and misguided, the concepts crystalized in the course of the research. I have not attempted to list the false starts, the dead-end trails, the quirks and the slip-ups, though they all happened. I am recounting my experiences not to demonstrate

Studies in Qualitative Methodology, Volume 2, pages 119-140.

ISBN: 1-55938-023-3

how the general rules of sociological method are distorted when applied to a "real" and social world and not to measure the difference between a methodological ideal and the messiness of human interaction in the research field. My reflections upon a piece of research are more focused. The research examined the experiences of married women and, as a woman researcher, I was pitched into an area where there had already been considerable debate.

Since Roberts (1981) published her influential collection of articles on doing feminist research, researching women has become a well-discussed area in sociological method. Feminist research has offered a broad critique of methodology and gender that was relevant to a whole range of issues encountered in the progress of the research project I undertook. It is these issues which provide the framework for my reflections. I was concerned with the conceptualization of women in sociological theory and research design, the ethics of establishing and conducting a program of research, the processes of interviewing women, generating rapport and humanizing the research experience and finally I was concerned with the role of research in defending the interests of women. Feminist research provided both a framework and a foil for my own research on women and was an invaluable guide to the area, although not one to be used uncritically. As a social movement, feminism has attempted to do more than make sound sense methodologically. It also has the political intent to express the interests of women and to articulate the nature of their structural and ideological subordination. Although acknowledging the wider aims of the movement, it was its implications for methodology and research practice that preoccupied me.

APPROACHING THE RESEARCH AREA

In 1984, I embarked upon a research project that was the basis for a Ph.D. thesis. I was in my late 30s and although I had taught sociology for a number of years I regretted never having had the time to devote to a substantive piece of research. I had had a break from teaching to look after young children, but had used a move to the Westcountry to re-enter academic life and to remedy what was for me an unfulfilled research ambition. On returning to sociology I was struck by the extent to which feminism had penetrated theorizing. This was particularly so in the area of marriage and the family. Here there was interesting and vibrant debate and this was the area in which I wanted to do research.

Much of the feminist critique had been aimed at "familialism" (Barrett and MacIntosh 1982) and conventional families (Oakley 1982). Living near a port town, with a large Royal Navy presence, there were many women whose husbands were regularly and frequently away and large estates of service personnel where this domestic situation was common. Husband-absent marriages seemed a relatively uncharted area and one ripe for exploration. As a part-time lecturer I was not in a position to apply directly for research funds, but colleagues were readily persuaded to apply for institutional support for the proposed project and I became a full-time research assistant. My research focus was on women married to Royal Navy personnel, women whose husbands were intermittently absent from home.

THINKING ABOUT MASCULINISM AND FEMINIST RESEARCHING

One major issue of the area seemed to be the extent to which theoretical frames and fieldwork processes, used in studying women, were "masculinist" in orientation. As a critique of research methods, feminist researching had advocated a greater reliance on qualitative techniques, had laid emphasis on women researching women and had brought a greater awareness of the power structures that impact on women's lives. But there seemed to be different ways in which masculinism could be interpreted and these had different implications for how the reserach was to be conducted. The term "masculinist" had been applied to survey methods, as they are associated with objectivity and rationality, attributes of public life and characteristics of men (Reinharz 1979; Morgan 1981). I considered this aspect of the critique less acceptable since it seemed to reify sexual divisions, delimiting intrinsically masculine and feminine traits and ways of apprehending the world. It took an active part in itself constructing gender and ran the danger of becoming an unwitting instrument in the "building (of) a methodological ghetto for women" (Graham 1983, p. 136), where feminist research can only use qualitative approaches. It also lacked a sense of history as it ignores the achievements of survey methods in detailing women's lives.

Seeing a role for surveys I had administered a postal questionnaire to a large sample group of women married to naval servicemen. It aimed to collect simple descriptive information, such as where women were living, how long they had been married, how much service separation they had experienced in recent years, if they had children, how many they had, their education and recent employment histories. These data later proved valuable in the contextualization of qualitative material derived from in-depth interviews. The information gained on numbers and trends was useful and this seemed a valid exercise without going the whole way toward positivist objectification of the population and the quasi-experimental manipulation of data.

Within the project, the term "masculinist" had a particular value and this value lay in the critical appraisal of past perspectives on the lives of women, including women with absent husbands. It was here that masculinism had an important bearing on the topic area. Intermittent husband absence was approached in the literature as an atypical marriage/occupational pattern and one that generated problems primarily for wives. It was never conceptualized as the husband's problem; he is a free; quasi-single individual whereas she is essentially dependent, part of somebody else. For him absences from home could be sad or exciting, but the reactions were superficial, idiosyncratic and without wider significance. Husband absence, by contrast, generated psycho-medical symptoms, frequently aggregated into syndromes, among wives (Isay 1968; Taylor et al. 1985). There were no assumed syndromes among absent husbands unless they are returning veterans (Cuber 1945; Bey 1972). Wives left without the daily presence of their husbands were seen to be living an abnormal lifestyle and one that was unanimously regarded as stressful. The empirical search had been for psychological disorder and disrupted patterns of eating and sleeping among wives, together with emotional disturbance among their children. Furthermore, analyses often contained assessments of the efficacy with which different types of wives "coped" with their "unnatural" home circumstances.

Models which tried to determine the levels, extent and nature of wifely distress contingent on husband absence qualified as "masculinist" in formulation since they contained a covert assessment of women and their adequacy as wives. It was a difficult task to acknowledge the structuring of women's lives by their husbands without reproducing a masculinist methodology. The assessment of women as wives is common in the approach of organizations to the wives of personnel and central to the conceptualization of "incorporation" (Finch 1983; Callan and Ardener, 1984). The Royal Navy was not exceptional in this and had developed an institutional view of the good and bad naval wife. The institutional modeling of naval wives was crucial to the study, their relationship to their husbands and to their husbands' work the centerpiece of the research. The women were picked because of the potentiality for their incorporation. However, while it was important conceptually to accept the social and structural implications of incorporation, it seemed equally important not to build such assumptions directly into the methodology. Wives were not to be conceptualized as only reactive to their husbands. It was a methodological essential that women were given the ontological status of being full social actresses, however structured their lives might be.

The feminist critique of methodology has been particularly pronounced where the area of debate has been that of marriage, wives, and domesticity. Although home and family had always been viewed as the main arena for women, feminism had made them more visible within it. Discussions of family life no longer revolved around the concept of "the couple" and there was recognition of the diverging interests and unequal powers of husbands and wives. But one problem remained. Women have largely supplied the data on family life. Although the theoretical frame had changed, the charge could still be made that a sociology of wives and not a sociology of marriages or the family was being described. The issue remained as the nature of marriage was being explored more through the experiences of women than men. Although I could be accused of compounding the partiality, a defense to this criticism was available in the work of Bernard (1982) in that I was offering a description of her marriage not his. There was also the caveat that marriage has a greater salience in women's lives; as women are socially "more married" than men, her marriage was then for me more analytically important than his. Yet I recognized that researching women as family spokespersons implicitly confirmed the domesticity of women and reproduced their gendering. The status of women in marriage and research, the extent to which this could be acknowledged in the methodology without simply reproducing the power relations of gender I regarded as difficult issues at the time and continue to do so.

Feminist researching had led to the reconceptualization of women in sociological theory. As it translated into fieldwork practice it had introduced fresh protocols and procedures and sponsored a new epistemology in stressing the value of women researching women. Although there was a superficial fit between these objectives and my research design, I recoiled from the assumptions of a "we-women" methodology, one which presumed a natural rapport and understanding embedded in gender. In collecting data it appeared an easy trap for the over-confident. It suggested a gnostic knowledge of other women's lives, and worse, it seemed to represent another ideological brick in the construction of the female personality and its intuitive sensitivity. It implied that gender was an unbridgeable gap in human understanding or concealed an essentialist argument that attempted to distinguish between understanding and true understanding. I was, therefore, reticent about claiming a special gendered place for me in my research into the lives of other women.

Nevertheless, whatever my attitude, the assumptions of mutuality were subtlety assumed in the project's inception. In part my request to undertake the research project was supported because it was felt that as a married woman with children I would be well suited to interviewing other women similarly situated. These assumptions were alluded to rather than clearly articulated, but are a measure of the extent to which the assumptions of feminist researching have penetrated research design.

The critique of masculinism was a potent influence on my approach to husband absence and in the initial stage of the research project I was very concerned with its varied implications and alert to its potential problems. I was anxious not to doff my hat toward feminist researching, to claim an affinity without taking it seriously or approaching it critically. These and other issues were to recur during the fieldwork and analysis stages of the research.

GAINING ACCESS

The women who were the subjects of the study were chosen because of their relationship to husbands and husbands' occupation. I wanted the women who were interviewed to represent a broad-based group, to be women of different ages, living in different areas and married to both officers and ratings. I hoped that this would ensure that the study would be of husband absence and not other factors. Other surveys in the area of service families had concentrated their efforts on service estates in port towns (Nicholson 1980; Stewart 1984). These studies included only one particular group of naval wives, those who were young, married to ratings, mobile and at a similar point in their life course. As it was interlocked with factors of class and age, the study was as much an analysis of these factors as issues of husband absence. The concentrated and more accessible community of service estates was unrepresentative of other types of naval wives as it excluded those living remote from naval bases, home-owners and underrepresented older women, with maturer families. I was keen that these other groups should, as far as possible, be included.

If the sample of navy wives was to be broad, it was essential that the cooperation of the naval authorities was secured since only through them could access to a widely based population of naval wives be obtained. The Royal Navy was approached to elicit its cooperation and support and to gain access to their personnel listings, from which would be derived a sample of wives. This was not some polite formality prior to being given a free hand, but a delicate series of discussions and negotiations. Within these negotiations a number of hurdles were crossed. Also the means of gaining access to women as potential interviewees structured other aspects of the project.

Research on service personnel inevitably encounters security problems. Therefore, it was hardly surprising that the Royal Navy was apprehensive abou any organization having access to personnel files. Access to such records wa limited, even within the Royal Navy, and they were certainly not for outsid eyes. There was an additional problem. The Ethics Committee of the Roya Navy had in the past developed regulations, it was claimed, to protect th civilian status of naval wives; they were not to be contacted by civilian or nava authorities without the prior permission of their husbands. Although the Nav was clearly interested in the consultative value of outside research, initially thes problems seemed to be major stumbling blocks. Eventually, however, compromise was reached and a listing of all the personnel in the administrativ region of Western Area was sent to the Family Services section of a local nav establishment. No names or addresses were permitted to be removed from thes premises, but all replies to a questionnaire survey and later invitations to a interview were returned to the Polytechnic. This means of contacting wome was cumbersome, but it protected their anonymity and fitted in with the Navy regulations on security.

This problem was complicated by another issue. The research could go ahead so long as all communications were addressed to both husband and wife, although they could contain letters that made it clear that wives were the group being researched. The Navy exhibited a concern for privacy tempered by a patriarchal view of wives. It was clear that the incorporated status of wives was not only part of the subject matter, it was also hard to resist its intrusion into the methodology of the area. The procedure made me feel uncomfortable but this compromise with naval regulations was more acceptable than the original suggestion of the naval authorities, that each husband's permission be sought for his wife to be contacted. This was the basis on which the research went ahead.

There were also other issues involved in these early negotiations. The naval authorities displayed a certain nervousness about the project's capacity to uncover a seamier side to naval life that would then be widely publicized and potentially sensationalized. Here I stood firm on academic probity and professional integrity. I pointed out that research was not an exercise in moral entrepreneurship, but would present them and the public with findings and interpretations that they would then be free to challenge. Methodology contains the public morality of research, and although an emphasis on protocol and procedure is an unfashionable aspect of methodology, for me it was important in the establishment of a research relationship with the powerful gatekeepers of information in the area. It was not only their political value, but also rules of protocol and procedure that seemed essential in the mechanics of research, where researchers have both a public place and the task of prying into the recesses of individuals' private lives.

Making contact with the women through the administration of the Navy meant that I already had a particular identity. This set in train what I initially regarded as another series of negotiations, whereby women interviewed had to be reassured that, although the survey had the Navy's blessing, I was independent, their cooperation voluntary and the information imparted entirely confidential. On meeting the women I often had to clarify my relationships, but negotiation was not the right word for what transpired, since this suggests that it was comparable to the dealings I had had with the naval authorities. The women needed some clarification and reassurance about who I was and what I was about. In replying to the invitation to be interviewed they had already agreed to cooperate in the research and, therefore, there was no element of persuasion in the discussion. Barnes (1979) sees researchers as people who inevitably cannot be entirely honest with any of the parties. Dishonesty may be too strong a word but there are undoubtedly elements of self-interest and manipulation in these negotiations and discussions, and the compromises that come out of them.

The means of access to the women structured the research in other ways. It contained implicit definitions of who were the married. Being married was operationalized through the naming of a wife as next-of-kin on the listings of personnel. This was in line with a common-sense definition of marriage that relied on self-identification. However, it became clear that service marriages were shaped by a greater weight of administrative constraint than most civilian counterparts. The Royal Navy uses a stringent definition of marriage in the regulation of pay and accommodation. For instance, to obtain married quarters servicemen must prove themselves legally married. As only 17% of the Royal Navy and Marine personnel had never lived in such accommodation (MOD 1983), this test of entitlement had at some time been applied to the vast majority. Also, other benefits and allowances are obtained only on the demonstration of legal matrimony. Therefore, a financial premium is available on marriage with the administrative upshot that legal matrimony is indirectly encouraged. This influenced the population of married women from which the sample was drawn.

INTERVIEWING WOMEN

I decided to collect detail on the marital histories of the women in a series of in-depth, tape-recorded interviews. They were loosely structured and organized around the topic areas of housing and neighboring relations, service separations, relationships with children, employment patterns, housework, household routines and household management practices. The questions formed a rough chronology from when they met their present husband, through past and present experiences to their anticipation of the future. I devised an interview schedule detailing questions but used it only in the first interview. The conversational basis of this interview was stilted and disjointed, largely because I was only half-listening to what was being said in my preoccupation with keeping to the schedule. In subsequent interviews I memorized a list of topic areas and had a list of questions to introduce new topic areas if the woman did not bring up the particular issue herself. These topic areas were then woven into the discussion in an order most appropriate to that woman's biography. This maintained the conversational flow and facilitated both more careful listening, the ability to return to past issues and events when new areas were broached, and helped me to spot and check any apparent contradictions. Another advantage was that as each woman's story was told I was able to pursue points using her words and expressions, her approach to the issue areas. Although the interviews covered the same ground, this approach enabled the women to emphasize and introduce experiences in the discussion areas that were important to them. The format of interviews then became more like prompted, episodic story-telling, rather than any clipped question and answer session.

A strong argument in feminist researching has been that, when interviewing women, the researcher should recognize their social situation, their relationship to domestic and wider power structures (Roberts 1981; Graham 1983). In particular, the difficulties of interviewing couples and the dominating behavior of men in couple interviews have been noted (Edgell 1980; Mason 1986). It was important that women should be interviewed when they were on their own. Although appointments were made to interview only the women, on two occasions husbands were present. His presence transformed the interview; he altered the questioning, the woman's answers and sometimes he joined in. Even when he did not speak he communicated what he felt by means of what has come to be known as body language and his reactions were monitored by the women in their replies. It was difficult to decide how this situation should be treated. At the time I decided not to abandon or cancel them because this would make a pointed issue and a re-appointment may have created difficulty between the woman and her husband. They have, however, not been discounted since the circumstances of all interviews are variable; each has its unique elements. It was a major problem to decide the extent to which the circumstances of the interview has varied so much that they are not the same event. Also, having interviewed someone, it was hard to blank out from one's mind what they had said, to undo the experience.

Although my intention to interview women when they were alone was on two occasions subverted by their husbands, there was also another, initially less apparent, aspect to this issue. Not only husbands were important in this context. In a further eleven interviews children were present. In other fieldwork accounts the presence of children appeared not to be noteworthy and was rarely recorded. However, the presence of children was not inconsequential, especially in the way that the interview was conducted, since interviewing a woman partially distracted by children presented its own problems. These interviews were on average much longer and I had to be more patient in repeating questions and ensuring that the women had fully said what she was going to say on each topic.

Another well-beaten path in debates on methodology concerned the identities individuals bring to the interview situation. With feminist researching there had been specific interest in women interviewing other women (Oakley 1981; Finch 1984; Cornwell 1984). These accounts had emphasized the ease of rapport, the empathy of gender and the dangers of "objectifying your sister." This approach to research emphasizes the potential for mutuality between the researcher and the researched. The emphasis on mutuality was particularly relevant where researchers shared a situation, or experience with the interviewee and that situation or experience was the focus of the research, as with Oakley on childbirth and Finch/Spedding on clergymen's wives. Being a woman/wife/mother gave me common ground with the women I interviewed, but no personal knowledge of the specific issue of marriage to a sailor; ignorance of this was, after all, what prompted the research.

These previous discussions of research methods had raised the vital issue of gender, but I was also aware that in the interviews with naval wives, gender was cross-cut by other identities. The women interviewed also placed me in terms of age and imputed level of experience. The most common questions were: Was I married? Was I married to a serviceman? And did I have children? The answers given supplied the biographical framework within which they could place me relative to themselves; it enabled them to contextualize their answers. Age differences altered the dynamics of the interview. Older women spoke more authoritatively, more confidently about their marital experiences. Their accounts were more reflective, more accepting of marriage to the Navy, full of what they had lived through as much as what they were living and their accounts were tied to the assumptions of an earlier generation. I also arrived with an occupational identity and there was often a request for this, and my relationship to the Navy, to be clarified at the beginning of the interview. Some assumed that I would have the power to alter naval practice and the commonest reaction was not wariness but forthrightness as many women treated the session as a rare opportunity to be heard.

There are issues of differences as well as commonality in the relations between women. Although rapport was established with all women interviewed, conversations were variable in their ease of flow, a product of the interface of the women's and my identity. Oakley discussed how she befriended four women selected into her sample on childbirth, but does not identify their characteristics, other than gender, as a background to the openness and the potentiality for befriending. There are limits to which women can befriend other women and a sensitivity to the limits of friendship and empathy are an essential ingredient of good research. I felt more comfortable with some women than others and this led me to wonder about the subtleties of rapport.

Nevertheless, perhaps there should be greater caution in overemphasizing the issue of gender in interviews. How, for example, was Parker's (1985) collection of edited transcripts of soldiers' wives talking different in nature from, or less adequate than, one collected by a female researcher? While accepting that gender made a difference, it remained unclear exactly what this difference was. There may be differences of fellow feeling and nuance in the dialogue of same sex/different sex interviews that do not affect the informational content or the data. And again there is the danger of reproducing feminity and female personality in claiming a special place for women interviewing women, without noting the specific intents and characteristics of the researcher.

RAPPORT AND EGALITARIANISM

Apart from identities, there had been concern with the processes of interviewing. Classic texts that offer recipes for the good interview stressed the importance of rapport. As Oakley (1981) notes, the work in this context did not denote the mutuality of a sympathetic relationship,

> but the acceptance by the interviewee of the interviewer's research goals and the interviewee's active search to help the interviewer in providing the relevant information (p. 35).

Rapport then meant an easy flow and truthfulness in their cooperation with the purposes of the research. In the project on naval wives, rapport was established easily in concurrence with the experiences of Finch, but this did not mean that "the model (of the interview) is, in effect, an easy intimate relationship between two women (Finch 1984, p. 74). The event was still an interview; the purposes of the situation were not transformed by gender. The discussion was structured, purposeful conversation and I was structuring and, therefore, controlling it. The women's questioning of me was minimal compared to my questioning of them and their questions were different from the questions I asked. They looked for perfunctory biographical detail or sought reassurance on the normality of their feelings and experiences compared to other women who had been interviewed; I did not seek reassurance from them. Although the women asked me questions, they were not as interested in me as I was in them; the interviews were conducted in their homes not mine; they offered me hospitality, I did not offer it to them; I recorded the words of the interview while none of the women that I interviewed were, to my knowledge, recording me. The women I interviewed were opening personal life to scrutiny and I was not. I could claim no special knowledge of what marriage to a sailor was like. I could never say "I know" in any personally authoritative way and this lack of intimate connection then led them to detail what it was like. Empathy has its limits and its drawbacks. It could shut down conversations as readily as it could stimulate them. Similarly, although the interviews were friendly they stopped well short of friendship. The emphasis on friendship and empathy seemed side-track issues in the phenomenology of the interview situation.

There was an element of hierarchy in the interviews with the women. A sympathetic demeanor, and an intent and willingness to answer questions was insufficient to term the interview egalitarian, since the relationship between interviewee and interviewer is never symmetrical. The feminist model of interviewer as friend or sister is not a description of the relationship but a political appeal to commonality among women, which may be a far cry from the actual relationships between women. Women do not confide in interviewers as they would a real sister but, if they do, as a sister in an anonymous and collective sense; the relationship is also not that of a friend, since friends are people with whom there are established and long-term connections, unstated understandings, and possibly appearances to keep up, positions to defend, and even axes to grind. Oakley (1981), despite her conceptualization of the interviewee as friend, appeared also to recognize the limitations when she argued that women are able to talk freely to interviewers because of their lack of close relationship,

> It was generally felt that husbands, mother, friends, etc. did not provide as sufficiently *sympathetic or interested* audience for a detailed recounting of the experiences and difficulties of becoming a mother (pp. 50-51, emphasis added).

The interviews worked because the interviewer was a sympathetic stranger. The process of interviewing may be like the early stages of befriending, but this is radically different from interviewing an existing friend. Oakley's model appeared one of possible and putative friendship and sisterhood, not the real thing.

Ease of communication was also not taken as the mark of egalitarianism in the interview situation. Equality is not the same as empathy; equality is not synonymous with ease of rapport as the openness of the women was their openness to manipulation and use by another

> ethical dilemmas are generic to all research. . . . But they are the greatest where there is the least social differences between the interviewee and the interviewer (Oakley 1981, p. 55).

Oakley and Finch are right to see the issue as an ethical one, not just a moral one, since ethics recognize the fundamental inequality of the situation and the need to protect those unable to protect themselves. Feminist researching should similarly accept the inegalitarian aspect of the interview and not attempt to conceal this in a plea for sisterhood. It is impossible to control the way in which published material is used; this is beyond ethics. But not deceiving or misleading people who are being interviewed and recognizing the vulnerability of women divulging personal information to another woman would seem to be the bottom line.

There were other qualities to the interviews besides those of a woman talking to women. Viewed in a different light the relationship could be compared with that between a counselor and the counseled. As the women were talked through their domestic histories since marriage, painful memories were stirred and accounts became infused with emotionality. A number were tearful, sorrowful in remembering the deaths of loved ones or in recounting unhappy episodes or situations. Although I gave them what sympathy and reassurance I thought necessary at the time and I tried to control my own wet-eyedness, there seemed to be few research guidelines for this situation. The traditional protocol of research is that of non-interference, to leave as you find, to finish the interview on a light note. However, social interaction leaves a wake and this is especially so when it entails self-examination, enabling people to draw new conclusions and hence to re-direct their lives. There was a therapeutic element in the discussions. This often made me feel drained and thoughtful while those interviewed invariably claimed to have enjoyed the discussion and to feel much better.

Counselors would not set out to rake through people's biographies and to pry into what may be tragic memories and human despair without some training and clear guidelines for their behavior, but researchers do. The similarities between interviewing and counseling reflect hierarchy in the situation as they may be asked for advice. I was taken as someone of some wisdom, if not authority, and I was taken aback by this assumption since I felt no more equipped to give advice than anyone else. The emphasis upon a shared humanity, a common phenomenology of women may have weakened the inhibitions of female researchers in commenting on other women's lives. But whatever the current efforts to link interviewer with the interviewed, I still felt uncomfortable and reticent when placed in this position.

HUMANIZING INTERVIEWS

The objectivity central to the traditions of scientific method has been seen as not only inappropriate but also dehumanizing in the conceptualization of the interviewee as passive respondent. Where subjectivity and the interactive processes are acknowledged, they are seen as an adjunct to eliciting cooperation and cajoling information, or an obstacle to detached and valid researching. The manipulation of the personal used to be approached as an issue of fine-line treading between the twin dangers of under- and over-rapport. However, the basis of the critique has changed as, instead of the discussion dwelling on the effectiveness of interviewing technique, it has attacked this style of interview as "de-humanizing" and, therefore, unethical.

"Humanity" has become something of a watchword in research and developed a special relationship with feminist researching. I had problems in interpreting this in any practical way. It seemed to have a variety of implications. At one level discussions of the "humanity" of research processes could refer to patterns of interaction that underlie all fieldwork encounters. Here interviewing was not some special behavior wherein the rules and processes of interaction became suspended. All interviews are mediated by human culture and conceptualization. Interviews proceed on common-sense understandings,

> all research is grounded because no researcher can separate herself from personhood and thus deriving second order constructs from experience (Stanley and Wise 1983, p. 361).

The text of the transcript is inevitably that of conversation. Alternatively, if you accept the perspective of Goffman (1956) on the interactions of everyday life, you are accepting one that has a fairly jaundiced view of humanity, since all presentations of self and episodes of interaction are manipulative.

Following this logic the research interview would be de-humanized if it were not as manipulative as other action. This is a view of humanity, warts and all, and one where humanizing research appears not such an attractive proposition.

All this is, however, sufficiently obvious that the argument for "humanizing" must have a different basis. I concluded that the acceptance of the humanity of the researched was an attempt to link the research act into a humanist philosophy, a wider philanthropy and in this particular context a feminist philanthropy. These arguments have a tremendous moral appeal. Few researchers would feel comfortable in describing their activities as the calculating and artful management of other people. Nevertheless I also thought this line of argument somewhat deceptive. A sense of fellow-feeling and identification with the putative interests of those whom you are interviewing is insufficient to change substantially the nature and events of the interview. The approach seemed part of a professional ideology that attempts to gild research, to make researchers appear virtuous and to link research to altruism. Researchers' claims that their research has due regard for the humanity of the researched, that it is non-manipulative, seemed part of their professional rhetoric. The feminist form of this rhetoric also contained the implication that feminist research did not exploit other women or that the women researchers in this fold were not ambitious or competitive careerists. Arguments for the humanization of research seemed an attempt to mystify the research process. They retained an asocial view of the fieldworker and were unhelpful guides to the events of the interview.

Methodology should emphasize the human and social nature of research. It is vitally involved in debates about how research is constructed, how the personal is woven into the collection and analysis of data and the consequences of this. Humanistic arguments are central to ethical position but are not a substitute for an honest appraisal of the events of the interview or the role of the fieldworker within it.

ANALYZING TRANSCRIPTS

Much of the recent methodological interest seemed to focus on data collection as the major problem area of research, but the analysis presented at least as many difficulties. Initially I listened to the tapes, but the manipulation of data required them to be in written form. I duly converted hours of tape into piles of text and in the process the data became more homogeneous. I avoided accent and the completed transcripts read with the same tone. I then began the purposeful reading of text, the search for patterns and attempted to link the experiences of people with their social situation. It was an attempt to discover common themes, variations and the possible sources of variation in the marriages and lifestyles of naval wives. The interviews all had a quasi-chronological format and the analysis took the form of life course analysis. It seemed the framework most suited for pointing up the developmental threads in biographical data, sensitive to maturational change and historical context. This enabled the varying experience of marriage to be related to varying circumstances. These analytic qualities gave it a number of advantages over the cameo portraits of case studies. One single woman could not speak for all or be picked up for her typicality. The object of analysis was also to find and relate other strands in the women's lives to their marriage patterns and it facilitated the analysis of movement through structure and within the relationship of marriage.

Tracing the women's life courses since marriage seemed the most apt way of bridging the gap between biographical detail and social structure. However, difficulties with the comparability of data had to be resolved. The sources of variability within and between the interviews were legion as each followed the biographical contours of the individual and there was an evolutionary element in the data collection, as the series of interviews progressed. Answers were pitched at different levels of generality and, although I "probed," probing did not always yield a set of comparable answers. Some women who were interviewed were like coiled springs, already preoccupied with the issues to be covered and eager to tell their stories, while others were more reticent. Some women gave accounts full of names, dates, events, places and personalities, while the answers of others were more clipped and detached from specific events.

Although interviews are not quasi-experimental situations, I was still concerned that there should be ways in which the adequacy of the material could be gauged in terms of the purposes of the research, the thoroughness of the interviews evaluated and the good interview distinguished from the bad. One way in which the interviews were assessed was in terms of their thorough coverage of the same ground. Issues of meaning and accuracy were more problematic, but, as far as possible, these were checked as the interviews progressed. I periodically recapped areas of discussion to make sure that I had understood the issues described, returned to contradictory statements and initiated discussion of the same area from a variety of standpoints. During the interview what people said was never directly challenged as research interviews do not take the stance of investigative journalism, relying more on the sympathetic ear and the assumption that relaxed individuals will recount their version of "the truth."

In order to pick out common themes in the marital experiences of the women who had been interviewed, sections of text were categorized under the broad topic areas that had originally structured the interviews. Each section was then examined for common experiences, attitudes and strategies and these behavioral and attitudinal items, as detailed by the women, were the basis of the analysis. Although qualitative in nature I approached the data in a highly methodical way. To check the commonality and thoroughness of these themes in the interviews, I constructed a chart relating biographical detail to experiences and attitudes. This was in the form of a large matrix, where personal details and information and the chosen topic areas were recorded against each woman to limit the risk of impressionistic analysis. On the chart, almost all boxes were filled and it recorded the type and extent of attitude and experience among the women interviewed.

Nevertheless, however methodical I was, this did not obviate the data's qualitative nature and the fact that analysis was always the search for uniformity among the unique. Each transcript was an historic document; when the experience of getting married for one woman, then aged 43 and married for 22 years, was compared with another, aged 20 and married for 3 years, different historic settings came into play. I had difficulty in distinguishing the historicity of the described events from the length of perspective from which they were viewed; younger wives seemed, for instance, more dissatisfied with married quarters and their furnishings than older women. This could be related to the higher aspirations of domestic materialism among the present generation of young brides or the flush of enthusiasm in "homemaking" that accompanies the early years of marriage and wanes with the passing years. Life course analysis was unable to resolve this. The ways in which women described their lives and the ways in which they answered questions was also variable. The levels of generality among the answers varied. Some transcripts were more story-like and fuller than others and it was hard to treat the taciturn account in the same way as the gushing; the fulsome were more likely to be the source of quotes and the unelaborated the basis of background collaboration.

During this process it was striking how often the women's accounts were variations on common themes and how, in the life stories and incidents recounted by sailor's wives, the same issues were confronted although different decisions may have been made. From these an attempt was made to construct typical pathways in their life courses and to detail the range of sentiments on particular issues. The initial search was for main issues and these stood out in broad profile. The early years of marriage were associated with geographical mobility and life in married quarters in contrast to the more settled existence but greater service separation of those in private housing found in later years. There were also the different neighborhood relations experienced by women in service accommodation and private housing, the independence learned in husband absence, the handing back of power in the readmittance of husbands into the home and the complexity of relationships with children. These themes were emphasized because they recurred and because they were frequently infused with emotionality. The broad brush strokes of the data were clear, but the detail, as essential in analysis, was not.

Feminist researching has brought a recognition of the complexity of women's lives and I felt that this complexity should be reflected fully in the analysis.

DEALING WITH VARIABILITY

Delimiting the broad themes and picking out major points was not a difficult task, but as analysis aimed to include all the points, attitudes and experiences, problems did emerge. To include all is difficult when transcript texts do not always show great consistency. There were differences in the stories told and there were contradictions within the same story. As variability and inconsistency were both evident it seemed important that they should be acknowledged and incorporated in the analysis if the data were not to be heavily bruised in their analysis. There was a wide range to, and ambiguity in, the expressed feelings of the women. When the breadth and variety of the women's experiences were included within a report of the findings, patterns were less clear and conclusions anything but definitive. Although not confused the inclusion of all aspects of what people say made analysis equivocal. Those who read the initial research reports said that they could detect no clear pattern or thrust of argument on the social situation of naval wives. One way to treat this issue was to weigh the statements into those of major and minor importance, although this suggested that some experiences were more valid than others.

Another way of dealing with the variability in data was to relate the differences to the particular circumstances of the women, to make the link between structural position and attitude. Unlike Parker (1985), and his interesting vox pop recordings of soldiers and their wives, this research was concerned not only with the statements of people, but an explanation and analysis of their situation. He was trying to describe the ranges of attitude and experience in editing his hours of tape recordings and pages of field notes, to offer a democracy of interpretation. The purpose of this research was to link the statements of the women interviewed with their social situation in a systematic fashion and to relate position to experience.

It was also a particular facility of life course analysis to trace connections between the different strands of a person's life. Here women who had left home prior to marriage had more positive attitudes toward husband absence. The youthfulness of wives was a factor and many of the older women described how they had painfully learned to conquer loneliness and develop vital friendships on which they could fall back when their husbands were away. But there were also different styles of marriage that women had developed for which it was not always possible to find a structural connection. Attitudes varied among the wives of officers and ratings, among women with and without children, and in different housing circumstances.

Examining the differences between transcripts and relating these differences to the life courses of the women was a basic process in the analysis; they were an obvious comparative component in the research. However, the variations did not stop there. There were contradictions and ambiguities within transcripts and these seemed a more intractable problem. Frequently women would make statements that would sharply contradict statements made earlier in the interview. Two clear examples of these were found when women were discussing their attitudes to husband absence and reunion, and the level of support they felt was forthcoming from other naval wives on the married quarter estates. Mixed emotions infused their descriptions of their husbands' departures and homecomings and contradiction filled their attitude toward their husbands' time at home and away. Women described the black despair of their husband's departure and the peace of time to themselves when he was not at home; they described the loneliness of his absence and the disruption of his return; they recounted their longing for their husband's return and the anti-climax of their reunion; they described their increased responsibility and independence and their greater vulnerability when things went wrong.

Women also gave inconsistent accounts of life in service accommodation. Many women characterized these estates as friendly and supportive. They emphasized the close links of women in the same metaphorical boat, the greater time women had to devote to friends when husbands were away, the chats over coffee, the shared excursions. In the next breath they would also describe their resistance to the intensity of close befriending, the pernicious gossip and cliquishness of close-drawn networks and social claustrophobia of life in married quarters.

These contradictions left me in something of a quandary. How were the different stories that issued from the mouths often of the same person to be treated? Could they be reconciled? Cornwell (1984) had noted similar inconsistencies when she repeatedly interviewed the same women in her study of health in the East End of London. She related these change to "changes in the relationship they had with me, the interviewer, and were also related to different techniques of interviewing" (p. 12). Cornwell employed the conceptual distinction between public and private accounts to explain these differences. Here "Public accounts are sets of meaning in common social currency that reproduce and legitimate the assumptions people take for granted about the nature of social reality," unlike private accounts that "spring directly from personal experience and from the thoughts and feelings accompanying it" (p. 15). This analysis may be helpful in distinguishing between types of statements, but was not one without its limitations.

This methodological assumption is part of a wider construction of public and private spheres of behavior, a construction that is part of the ideological fabric of society. In this context, it implies that statements are not of the same order, in that some are indicative of true feelings while others are a gloss on experience, that some comments are more real and less superficial than others. Cornwell's use of the division suggests that some statements are more socially shaped than others and it is part of the researcher's task to distinguish between the two. I felt unhappy about the imposition of the distinction. The division raised difficult epistemological questions about how I, the researcher, could and should separate a public from a private account. The distinction also assumed that a better truth could be found by peeling away the layers of social discourse.

The complexity of the women's attitudes and the ambiguity of their feelings was striking and the complexity could not easily be resolved by slotting data into the categories of public and private accounts. Marriage generates mixed emotions. Husbands were described as sources of comfort and support as well as anger and irritation and this genuine ambiguity seemed best preserved intact in the analysis. For many naval wives power and responsibility were uncomfortable sides of the same coin. Husbands absence permitted them to make more decisions and for some such decision making was burdensome. Neighborhood ties were both supportive and destructive. Both attitudes were valid and indicative again of the duality and contradictory emotions that are part of intimate relationships. I was concerned that in the analysis it should be recognized that intimate relationships are many-sided. Having been a vehicle for making women visible in methodology, feminist researching should also be the agent for detailing the complexity of their social experiences.

However, an allied concept of "public morality" was central to my understanding of the transcripts. I used it not as a format with which to reconcile the differing statements of women, but as a frame of meaning that influenced all their statements. All family forms and types of communal living exist in the shadow of "conventional families" (Oakley 1982). This shadow renders any alternative form odd, inadequate or experimental in nature. Throughout the discussions women contrasted themselves with what they believed "ordinary" wives and "ordinary" marriages to be. The concept pervaded their thinking. Here, Voysey's (1975) use of the concept "public morality" was more pertinent. Conventional and unseparated marriages represented an ever-present frame against which to assess their own circumstances. As the parents interviewed by Voysey could discuss their disabled children only within the context of everyday models of parenting, so the women I interviewed conceptualized and judged their marriage within the assumptions of marriage in general and all that this entailed. Sometimes their idealization of normal marriage and conventional families fed false comparisons and supplied them with a hook on which to hang all their marital problems. It was assumed by some that things would all be different and troubles dissolved by a "normal" and unseparated family life and the descriptions of these women often included an "if only" line of argument.

Others were at pains to describe how a naval way of life had prevented them from having an ordinary marriage, or how, despite the exigencies of service at sea, theirs was a normal marriage.

THE INTERESTS OF WOMEN

Many of the methodological issues discussed are concerned with the impact of the research on the women interviewed. But the research has also had an impact on me. I have developed a fierce identification with the preoccupations of naval wives and have found myself defending them against any criticism. The critique of masculinism in methodology has stressed the importance of describing and analyzing the lives of women in terms of the interests of women. For me this raised difficult questions about the role of the researcher in familiar debates on structure, consciousness, interest. Although I felt quite clearly that I was on their side, I was not quite sure what their side was. This problem emerged with women who shared the same social situation but who interpreted their situations entirely differently. Also feminist research dealt uneasily with the statements of women that did not fit the established critique of marriage. If you assume that there is a side that you can be on then the statements that do not fit can be dismissed or patronized as the words of the unaware or passed over as the empty cant of public morality. In the hierarchy of credibility they lose out.

This was not just a personal issue; in writing a report to the Navy on the findings I felt I must represent their interests, despite their diversity of opinion and without giving the impression that naval wives constituted a prima facie welfare problem, which they did not. In 1971 Becker asked sociologists "whose side are we on?" in his critique of value-freedom. Becker's arguments, so appealing when I read them a decade and a half ago, now seemed specious and paternalistic. They assumed that you could detect what the sides were even if those researched did not speak with one voice. It seemed to be one thing to know whose side you are on and quite another to know exactly what the interests of the side were and how they were to be advanced.

CONCLUSION

Becker described methodology as "a proselytizing speciality," since there was a "strong propensity of methodologists to preach a 'right way' to do things, because of their desire to convert others to proper styles of work, because of their relative intolerance of 'error'" (1971, p. 4). Becker also saw methodological debates as prone to fadishness and fashionability, preocccupied with some methodological issues and not others. His critique was aimed at a sociology dominated by quantitative methods. However, his remarks have a wider application. I had entered the research field in a much discussed area, that of researching the domestic lives of married women. Here there was more conventional wisdom and greater pressure to do the right thing. At times I was worried that I was swimming against the tide of opinion. Given the weight of moral pressure that infuses feminist researching, it was tempting to subscribe publicly but privately retain my reservations. This uneasiness about what I was doing in the research and what the literature suggested I should be doing has informed this article.

The critique of approaches to women in contemporary research was valuable in its assessment of theoretical frameworks and in its sensitivity to the conceptualization of womanhood in masculinist terms and the variability and intricacy of women's lives. It also suggested that there was a residual problem of researching women incorporated in the work patterns of their husbands, yet using a methodology in which they could be more than an adjunct to their husbands' occupation. My reservations concerned the framing of research processes which made a special case for women researching women. These seemed both to reify and mythologize gender and I remained skeptical of the extent to which the mechanics of information-gathering can be "humanized." The best way to "humanize" research is to recognize fully its place in social structure and human interaction and for researchers to be more aware of their manipulative and authoritative place within interviewing and analysis.

REFERENCES

Barnes, J.A.
1979 *Who Should Know What?* Harmondsworth: Penguin.
Barrett, M. and McIntosh, M.
1982 *The Anti-social Family.* London: Verso.
Becker, H.S.
1971 *Sociological Work: Method and Substance.* London: Allen Lane.
Bell, C. and Newby, H. (eds.).
1977 *Doing Sociological Research.* London: Allen and Unwin.
Bernard, J.
1982 *The Future of Marriage.* New Haven: Yale University Press.
Bey, D.R.
1972 "The Returning Veteran Syndrome." *Medical Insights* 4:42-45, 49.
Burgess, R.G. (ed.)
1984 *The Research Process in Educational Settings: Ten Case Studies.* London: Falmer.
Callan, H. and Ardener, S.
1984 *The Incorporated Wife.* London: Croom Helm.
Cornwell, J.
1984 *Hard-Earned Lives: Accounts of Health and Illness from East London.* London: Tavistock.
Cuber, J.F.
1945 "Family Adjustment of Veterans." *Marriage and Family Living* 7:28-30.

Edgell, S.
1980 *Middle Class Couples: A Study of Segregation, Domination and Inequality in Marriage.* London: George Allen and Unwin.

Finch, J.
1983 *Married to the Job: Wives Incorporation in Men's Work.* London: Allen and Unwin.
1984 "It's Good to have Someone to Talk to: The Ethics and Politics of Interviewing Women." In C. Bell and H. Roberts (ed.), *Social Researching.* London: Routledge and Kegan Paul.

Goffman, E.
1956 *The Presentation of Self in Everyday Life.* Garden City, NY: Doubleday.

Graham, H.
1983 "Do Her Answers Fit His Questions? Women and the Survey Methods." In E. Gamarnikow et al. (eds.), *The Public and the Private.* London: Heinemann.

Isay, I.
1968 "The Submariners' Wives Syndrome." *Psychiatric Quarterly* 42:647-652.

Mason, J.
1986 "Gender Inequality in Long Term Marriage; The Negotiation and Renegotiation of Domestic and Social Organisation by Married Couples aged 50-70." B.S.A. Loughborough Conference.

M.O.D.
1983 *Report on the Armed Forces Accommodation and Family Education Survey.* D/Stats (s).

Morgan, D.
1981 "Men, Masculinity and the Process of Sociological Enquiry." In H. Roberts (ed.), *Doing Feminist Research,* London: Routledge and Kegan Paul.

Nicholson, P.J.
1980 *Goodbye Sailor: The Importance of Friendship in Family, Mobility and Separation.* Inverness: Northpress Ltd.

Oakley, A.
1981 "Interviewing Women: A Contradiction in Terms." In H. Roberts (ed.), *Doing Feminist Research.* London: Routledge and Kegan Paul.
1982 "Conventional Families." In R.N. Rapoport, M.P. Fogarty, and R. Rapoport (eds.), *Families in Britain.* London: Routledge and Kegan Paul.

Parker, T.
1985 *Soldier, Soldier.* London: Heinemann.

Reinharz, S.
1979 *On Becoming a Social Scientist.* San Francisco: Jossey-Bass.

Roberts, H. (ed.)
1981 *Doing Feminist Research.* London: Routledge and Kegan Paul.

Spedding, J.V.
1975 *Wives of the Clergy.* Unpublished Ph.D. thesis, University of Bradford.

Stanley, L. and Wise, S.
1983 *Breaking Out: Feminist Consciousness and Feminist Research.* London: Routledge and Kegan Paul.

Stewart, C.E.
1984 *Clyde Submarine Base: The Effects on the Family Unit and its Relationship within a Service Environment.* Ph.D. thesis, University of Strathclyde.

Taylor, R., Morrice, K., Clark, D., and McCann, K.
1985 "The Psycho-Social Consequences of Intermittent Husband Absence: An Epidemiological Study." *Social Science and Medicine* 20(9):877-885.

Voysey, M.
1975 *A Constant Burden.* London: Routledge and Kegan Paul.

PALE SHADOWS FOR POLICY: REFLECTIONS ON THE GREENWICH OPEN SPACE PROJECT

Jacquelin Burgess, Barrie Goldsmith, and Carolyn Harrison

The Greenwich Open Space Project is concerned with social meanings and values for nature, landscape and the urban green. The research is innovative in that it has adopted the theory of Group Analysis and adapted therapeutic practice to run in-depth discussion groups with members of the general public. Geographers and ecologists acknowledge that the approach gives insight into people's environmental values and attachments which has not been achieved before, recognizing a level of authenticity which challenges both conventional theories of landscape and nature and the professional values of those involved in shaping public policy. Persuading policymakers to accept the legitimacy and representativeness of the research is a political process which has been more difficult because our arguments are based on qualitative rather than

Studies in Qualitative Methodology, Volume 2, pages 141-167.

ISBN: 1-55938-023-3

quantitative methods of social survey. Our experiences with the Countryside Commission over the last three years illustrate the extent to which it has been possible to persuade a policy-making community of the value of qualitative research.

The two year project was funded by the Economic and Social Research Council (ESRC) (£21,000), with £6,000 in additional support from the Countryside Commission for England and Wales. Carolyn Harrison, a biogeographer was the principal investigator. Jacquelin Burgess, a social geographer and Barrie Goldsmith, an ecologist in the Biology Department at University College London (UCL) were the two other staff members. Melanie Limb, a graduate student completing a thesis on community involvement in recreation provision and management, was appointed as research officer for the project. We wanted to develop a new methodology which could reveal the value of open land for urban residents and planned to carry out three phases of field research: small group discussions; a questionnaire survey that would incorporate the findings from the groups; and interviews with local authority officers to establish what professional values were being used to justify the conservation or loss of open land in the city. We expected the questionnaire survey to form the major piece of empirical work and to conduct its analysis using conventional quantitative methods. In the event, the in-depth discussion groups assumed major importance and represent the most significant contribution of the project in both academic and policy terms.

In this chapter, we discuss the genesis of the project, exploring our original aims and objectives, the academic contexts within which we worked and the demands of the external funding agencies. Then we focus on the research process itself, showing how the field experiences influenced our plans and modified our actions. We discuss the development of our interpretive strategy for handling the material from the in-depth discussions and the relationship between the groups and the questionnaire survey. Finally, we describe our relationship with the Countryside Commission and show how the research is being used in an active policy debate about countryside recreation.

THE GENESIS OF THE OPEN SPACE PROJECT

Theory, field experiences and policy are entangled in complex ways which, on reflection, are difficult to separate. The fundamental problem, as a recent collection of papers makes clear (Penning-Rowsell and Lowenthal 1986), is that there is no clear theoretical basis for understanding environmental values; they are essentially contestable structures of feeling. Some authors argue that environmental values may be traced back to earlier evolutionary periods and can be explained by various neurological and behavioral mechanisms essential to survival. Others feel that landscape values can be determined through experimental studies of psychological processes of perception and cognition, and reduced to stimulus-response models of behavior. A third group of authors draw on social theory, particularly cultural materialism, to interpret changing landscape values in terms of the relations between dominant and subordinate groups in different social formations (see also Cosgrove 1984). Meanwhile, nature conservationists continue to work with theories of ecological diversity that emphasize the biological bases of landscape features and stress the scientific and educational importance of environmental values (Usher 1986).

Some of these theoretical differences, notably those between culturally-based theories of landscape values and those of natural and social science, existed in our project team and the ways in which they were or were not reconciled had a direct bearing on the research. It is not common practice for the individual members of a multidisciplinary research team to discuss their internal dynamics in a public forum. We were able to externalize the workings of the open space groups through a mode of conducting their proceedings, a verbal and written record of their discussions, and detailed interpretations of their interactions. Other academics too, can appraise their conduct and content as the discussions are in the public domain.[1] But our internal dynamics remain hidden from similar scrutiny and so, in this chapter, we highlight our own leadership battles, alliances, value conflicts and taken-for-granted assumptions about the nature of research, data analysis and interpretation—all of which have influenced the direction and outcome of the project.

The Disciplinary Context

Exploration and fieldwork constitute the core of the geographical method and the discipline has a rich history of empirical research. The narratives of explorers in the nineteenth century; the historical interpretations of landscape development and regional studies of the early twentieth century; and the writings of the American cultural geographers provide a tradition as rich as the ethnographies of anthropology or the Chicago school of sociology. In common with other social sciences during the postwar period, human geographers turned away from interpretive approaches to embrace a positivistic philosophy, developing increasingly sophisticated statistical techniques to explain spatial patterns and processes. Over the last fifteen years or so, the hegemony of the "quantitative revolution" has been fundamentally challenged by other perspectives, notably those with a humanistic orientation that draw on the philosophies of meaning, and those in social geography that prefer marxist or structuration theories. Qualitative methodologies are re-emerging in empirical research and the special field skills of geographers are to be found in detailed observations and interpretations of landscapes and places (see, for example, Meinig 1979; Jackson 1984). Other qualitative approaches currently being developed are those common to anthropology, sociology, and the humanities: participant-observation; semi-structured and depth interviews; the interpretation of personal documents, archives and other documentary materials, including written and visual texts (see Eyles and Smith 1988).

Geography is an academic discipline that straddles the divide between the natural and the social sciences. Within the department at UCL, we have colleagues researching issues as disparate as recent earth movements and plate tectonics; paleoecology and acid rain; the analysis of economic and social problems in a wide variety of first, second, and third world countries; historical and contemporary landscape studies. We share a common cause that is best described as an endeavor to understand the complex and diverse relations between human societies and the physical environment. The departmental ethos is one which seeks to integrate the skills and perspectives of physical and human geographers. For that reason, landscape studies and the conservation of both natural and built environments provide a common focus for research and teaching. The disciplinary and institutional context accounts for our multidisciplinary team. Harrison and Goldsmith have cooperated for several years in teaching the Masters course in Conservation at UCL. Departmental research seminars persistently raise questions about the psychological, social and political dimensions of environmental management and conservation that cannot be answered by physical scientists alone. Collaborative work with social geographers is needed if new, culturally-based definitions of resources and environmental problems are to be forged (Warren and Harrison 1984).

Personal Motivations and Policy Contexts

Carolyn Harrison writes first about her motivations for undertaking the research.

In the early 1980s, a number of nature conservationists and landscape preservationists began to express the view that neither traditional scientific reasons for nature conservation nor exceptional justifications for landscape protection had served either cause very effectively (Green 1981; Shoard 1980). Mounting evidence of the loss of wildlife habitats and the extent of landscape change stimulated the search for new justifications for the extension of controls over the wider countryside in pursuit of conservation objectives. Conservationists re-emphasized the personal, social, and cultural benefits that come from regular contact with the natural world. Concern was also expressed for the unofficial countryside found in towns and waste places—the landscapes accessible to most people. As Mabey argued in *The Common Ground,* the viewpoint most commonly overlooked in the public debate about nature conservation was that of "the ordinary 'consumer' of wildlife" (Mabey 1980, p. 13). I wanted to focus on the attitudes and values of a cross section of people living in one London borough in a modest attempt to redress this neglect.

I was also interested in recreational behavior and the role of open land in peoples' daily lives, having recently finished a project which explored the recreational role of the urban fringe around London. Our questionnaire surveys of visitors to public open spaces in the Green Belt revealed the highly localized catchments from which most visitors to a site were drawn and the apparent "failure" of sites in the countryside of the urban fringe to attract people from inner London (Harrison 1983). But I felt that a full explanation of the patterns we had identified was likely to be found in the nature of recreation itself and not just in the properties of particular sites. I wanted to do some new research in order to address these wider recreational issues and to develop a methodology more appropriate for investigating them.

The debate about the purposes and practices of nature conservation and the recreational role of sites in the countryside impinge directly on several areas of public policy, most notably those pursued by the Countryside Commission. The Commission has primary responsibility in England and Wales for protecting the countryside and promoting public access and enjoyment of it. The team approached them with the object of securing sponsorship of the Greenwich research during 1984, at the time when the Recreation and Access Branch of the Commission was about to embark on a major review of its countryside recreation policies. The Commission does not generally solicit requests for research funding but officers considered our proposal and we discussed how the research could be modified to accommodate their requirements. These discussions reaffirmed our commitment to use small, in-depth discussion groups as the first stage of the project. It was also agreed that the findings from the groups would be used to design a questionnaire for a household survey. The Commission expressed no interest in the third phase of the research. We were pleased with the outcome of these discussions because it enabled us to approach ESRC for funding, with additional financial and political support that demonstrated the relevance of the research for conservation and recreation policies.

Our decision to work in Greenwich and our interest in the open space policies of the local authority arose because Goldsmith and I were already actively opposing the East London River Crossing (ELRC), a major development by the Department of Transport that threatened several open spaces in the borough. Our involvement with the public enquiry into the ELRC had a material bearing on the kind of methodology we proposed to use in the new research. Critically, I undertook an analysis of the impact of the road on the recreational role of Oxleas Wood, an ancient woodland of great ecological and recreational significance. In assembling evidence to be presented on behalf of Greenwich Council, we carried out a questionnaire survey of visitors to Oxleas Woods and among households within walking distance of the site. Doing the survey convinced me of the shortcomings of questionnaires. However carefully we recorded responses to questions about how the road would affect use of the wood, none adequately captured people's profound sense of loss; feelings of love and attachment for this landscape became little more than banal, clichéd expressions which conveyed little meaning. We used the survey results at the public inquiry, but we began seriously to question whether it was possible to devise more appropriate ways of enabling people to speak for themselves. Above all else, it was this field experience that convinced us of the need to pursue alternative methods of inquiry. We sought a method that would give full weight to people's own words, feelings and the symbolic meanings they invest in places. We began to consider small groups but the decision to use in-depth groups, drawing on the theory and practice of Group Analysis, stemmed from involving Jacquelin Burgess in our work.

Jacquelin Burgess writes about her contribution to the project proposal.

My research interests have focused primarily on the social meanings of places and people's attachments to locality rather than nature conservation and recreation. I had become disillusioned with questionnaire surveys in consequence of my work as a postgraduate. I had researched place imagery using a questionnaire survey that contained a variety of psychometric techniques to measure meaning (Burgess 1978). I found the methodology totally unsatisfactory, barely scratching the surface of deeply felt local attachments to place and unable to situate experiences within either local or national cultural contexts. Between 1978 and 1984, I explored a variety of qualitative methods: participant-observation in Billingsgate Fish Market and a study of the traders' relocation to the Isle of Dogs grew into research on relocation decision making and the impacts of place advertising (Burgess and Wood 1988); a longitudinal, oral history project in a mining village in County Durham with first year students; what might be best described as "investigative journalism" in researching the sense of place in the Fens for a television documentary (Burgess 1982); and developing interests in media texts and methods of textual analysis and interpretation (Burgess 1985).

My specific interest in groups had developed from a rather different strand in my life, however. I first became interested in group psychotherapy while a graduate student at Hull. Alan Ingleby, the student counsellor, allowed me to attend some of his student groups as a participant-observer. My interests in counselling developed at UCL: shortly after arriving in 1975, I made friends with Wyn Bramley who was then student counsellor in the College. She ran a staff group in which I participated for a year, as well as a student group in which I acted again as participant-observer. At this time, I was seriously considering abandoning an academic career to train as a counsellor. In response to that uncertainty and a personal crisis, Bramley encouraged me to consult Dr. Malcolm Pines, Director of the Institute of Group Analysis, and I spent 2½ years as a member of one of his therapeutic groups. The seeds of my contribution to the Greenwich project were sown in these groups. What struck me most forcibly were the ways in which groups establish a life and identity of their own; how members communicate with each other and the group as a whole; the storytelling; the ability of individual members to express feelings that others could feel but not articulate; how the moods of individual sessions varied; the power of feelings and the ability of people to express them in incredibly rich and figurative language. And what stayed with me, niggling away in my own subconscious, were the landscape symbols and metaphors, the ways in which individuals projected their emotions onto different physical settings. Environmental symbols continually appeared in people's descriptions of their dreams while everyday expressions, like "feeling lost" would take on immense psychological and geographical significance as someone might describe their feelings of desolation in terms of wandering across a vast empty plain, full of dark holes and threatening shapes, with a bitter wind howling in their face.

But is was difficult to see how this could become part of an agenda for geographical research. Between 1981 and 1983 I began reading psychoanalytic literature, as part of an attempt to understand what view of the self was being represented in geographical work and, at this stage, Harrison approached me with a preliminary outline of a project to look at people's values for open land in the urban fringe. Would I be interested in participating and did I have any ideas about possible methodologies? I was enthusiastic about doing a major piece of empirical research, excited about the possibilities of working in a research team, and keen to establish closer links with the only other woman academic in the department. Fundamentally, the project offered me the chance to develop a group-based methodology that could explore people's feelings and attachments for environments and places. We might be able to prove, once and for all, that "ordinary" people could talk about feelings for place; that they did not have "inchoate" experiences; that the essentially sterile responses to a variety of quantitative methods of evaluating landscapes were the fault of the method rather than any true reflection of the feelings of the participants. We knew of only two academic studies in our fields that had employed single group interviews (Mostyn 1979; Little 1975), but we felt that such groups could do little more than tap the most superficial level of feelings for environment. Crucially, we wanted to develop a methodology that would give people enough time and space to explore their emotions and attachments, in a social setting in which they would feel comfortable and secure. I felt that it might be possible to adapt Group-Analytic techniques for a field study. Harrison was prepared to take the gamble, largely because we had both been members of a task-oriented, staff group run by Bramley as part of a major curriculum review in the department (see Bramley and Wood 1982).

Barrie Goldsmith writes about his motivations for undertaking the research.

My participation in the project was based upon my interest in nature conservation spanning more than twenty years and involving responsibility for the M.Sc. Course in Conservation at University College London. Parallel lines of inquiry that I have pursued involve ecological or conservation evaluation, which is a controversial subject, and the selection of "indicators" to represent areas of high nature conservation value. During these studies, I have been aware of the need to know what people think are the important attributes of semi-natural areas, the criteria that we use to select protected areas, including people-centered criteria (Goldsmith 1983, 1990).

Another research interest has been countryside recreation and the study of the ecological impacts of recreation. It appears to me to be important that we find out what people are really looking for when they decide to go out into the countryside as this will help resource managers to balance the demand/supply equation and enable countryside staff in local authorities to provide for people while safeguarding the resource. A proportion of my research has been of a qualitative nature including the preparation of management plans and some work on conservation evaluation.

My personal interests reflect a lifelong love of the outdoors and what most people would consider beautiful places. When I left school in suburban south London, I chose to go to University at Bangor in North Wales. This was perceived as being beyond the last frontier by many people but was to me a major decision that I never regretted. I walked the Welsh hills every Sunday and easily slipped into an interest in field-based ecology. Rather suprisingly, since then I have been based in central, urban London but I live 25 miles away and prefer to commute rather than lose touch with a semi-rural environment.

My training has been scientific but I have always been suspicious of a lot of research that involves massive amounts of data, computer analyses, and so forth, and that at the end of the day may tell us nothing. At the initiation of this project I was hoping for some other approach that would tell us something about the values that ordinary people hold about the outdoors, open space, countryside, beautiful views and fascinating things such as orchids and butterflies.

We all embarked on the project hoping that we would be able to identify popular values for the countryside. We all had different ideas as to what these were likely to be and how they would be described and communicated to others. Maybe I placed more emphasis on the importance of communicating these ideas to people with a responsibility for the countryside, such as the National Trust, the Countryside Commissions, Nature Conservancy Council, and County Wildlife Trusts. My aims were:

1. To understand people's *perception of the countryside.* From the beginning I expected that small group discussions would probably be a forerunner to wider discussions or more extensive survey.
2. That discussions would need to be *structured* in some way, perhaps by geographical area, types of habitat visited, or age of participant. The kind of study that I expected was for us to refer people to an ideal environment, invite them to describe it, compare it with one that they knew well, liked a lot or visited frequently and to use this as a basis for investigation.
3. That the research would *help to explain* observations that many people have commented on, for example, why half a million people will pay to support a society that looks after birds while very few support a society dealing with butterflies. I also hoped that it would identify priorities for management. If, for example, we knew that a lot of primroses and bluebells were important to people it would encourage managers to coppice their broadleaved woodlands. If that were not the case it would be cheaper to neglect them and let them develop as wilderness.
4. That the results would be *representative* and *reliable.* Scientists present their results in the form of means and standard errors. The size of the latter is a function of the inherent variability of the material examined and the sample size. Scientists prefer to have small standard errors and thereby small confidence limits. I was not looking for rigorous statistical analysis but did expect social scientists to, at least, discuss the reliability and representiveness of their results.
5. That the results of the study would be *applicable* and *relevant* to a variety of countryside management agencies such as those mentioned above. I hoped that they would be in a form that could be easily communicated to others and that managers would appreciate the worth of what was done.

THE RESEARCH PROCESS

The project emerged from these rather different strands. Significantly, we had each assumed that the other members of the team understood our individual positions in terms of the theoretical basis for researching environmental meanings and values. We had assumed that we had a common perspective and had not clarified our different theoretical positions before the project began. Over the first few months of the project, it became clear that there were major differences between us in terms of whether we wished to adopt an approach that emphasized individuals' perceptions of open space and countryside or one that focused on the social meanings of different environments as they are expressed through discourse. Most empirical work, based in environmental psychology, was using questionnaires, psychometric techniques and photographic surrogates to explore perceptions and preferences for different landscapes. The possibilities of using qualitative approaches to explore different environmental discourses were not being discussed in the literature.

Our decision to use in-depth discussion groups raised major difficulties. It is important to stress that the approach was novel and experimental. We were pursuing a hunch. There was no published research on which to base our fieldwork. We knew that once-only groups were used in market research but no one had attempted in-depth discussion groups with members of the general public. The central question facing us was how to operationalize the groups in the field? The theory and practice of Group Analysis is well established (Foulkes 1975; Pines 1983) and the Institute of Group Analysis publishes its own journal. But in these circumstances, the therapeutic purpose is clear to all the participants. A psychotherapist will select patients for a therapeutic group; members will meet in the controlled setting of the consulting room; the group has a long life; and patients attend with the expectation and motivation that aspects of their lives will be eased through therapy. When group-analytic methods have been used in educational contexts too, there is a clearly defined membership, the group meets within an institutional setting and has clearly defined goals (Abercrombie 1983; Bramley 1979).

We faced very basic problems: Where to run our groups? With whom? How many groups? For how long? With how many people? How to find members for the groups? What should we talk about? How to conduct groups where members do not have the commitment to keep coming because it is part of a healing process or because they are gaining professional or personal insights? We were acutely conscious of the ethical issues raised by the research. By its very nature, being in a group encourages individuals to become intimate with one another and to reveal aspects of their lives and feelings that they would not do in other circumstances. Psychological and personal problems might be raised in the discussions which could cause harm if potential members were not carefully screened and if the groups were not led with a proper level of professionalism. We were, therefore, very pleased when our first paper on the project was accepted for the journal *Group Analysis*. It was recognized that we had understood the demands of running the groups in accordance with group-analytic principles (see Burgess 1986 and the discussion by Bramley 1986, p. 244).

Making a Start

We began in January 1985 under strong pressure from the Countryside Commission because the start had been delayed by ESRC. The Commission wanted results from the group discussions as quickly as possible. We held weekly team meetings during January and early February to decide our procedures for recruiting and running the groups. At the same time we made contacts with Local Authority Officers in Greenwich and began to familiarize ourselves with the borough. Key decisions were taken in our very first team meeting which Goldsmith was unable to attend. On the basis of her experience, Burgess proposed that the life of the discussion groups should not exceed six weeks: it would give people enough time to get to know one another and establish strong relationships without encouraging too much intimacy. But she was by no means sure whether people could actually talk about open space for that length of time. Adopting the standard practice for therapeutic groups, the length of each meeting was set at 1½ hours, and all were to start and end on time. Membership was set at between 8-12 people. We agreed that 4 groups would be the maximum we could manage, given the time constraints and proposed a timetable which, with hindsight, was wildly unrealistic. We would begin the first group on Feburary 25, and then recruit and run the other three groups in rapid succession. We expected to run two groups a week for most of the sequence. None of us realized the difficulties we would have in recruiting groups, especially among the Asian community and in Thamesmead.

In this first team meeting, we also agreed on a strategy for recruiting members and the constitution of the groups. We wanted to recruit people who lived in localities with different qualities and quantities of open space; we were anxious to recruit groups which were culturally homogeneous, with members from the same social class, as far as possible. We would prefer adults rather than adolescents, and mixed gender groups. Since we were most anxious to work with non-environmentalists, we decided to contact people in three ways; first, through voluntary groups such as tenants' associations, playgroups and local societies; second, through informal contacts made in pubs, clubs and community centers; and third, through the local media such as the Greenwich Council free newspaper. In the event, we took the first two avenues, writing to a number of community groups in the localities and then meeting people who were introduced to us. We used friendship networks to make additional contacts. On reflection, it was a pretty desperate process, not made any easier by the need to get the groups started (see Burgess, Limb, and Harrison 1988a).

Emerging Conflicts over Methodology

But what should the groups talk about? Goldsmith came to the second team meeting on January 16, and the tensions between scientific ways of thinking and organizing research, and qualitative, interpretive approaches became clear. There was considerable disagreement about the extent to which people should be directed in their discussions. Burgess proposed a sequence of very broad topics for the sessions. The first meeting should enable people to get to know one another and decide what they wanted to do in the groups. The second should focus on people's experiences of living in their area, what they do in open spaces and what changes they would like to see, while the third would concentrate on things they like and value in opens spaces. The fourth session Burgess suggested might consider people's dislikes and places in which they feel uncomfortable and frightened. That would be followed by a session on management of open spaces in the borough and the sequence would end with a final session, where it was very important to allow people to say goodbye to one another and to review the work of the group.

Such an approach was far too free and unstructured for Goldsmith who preferred a natural resource-based focus and controlled approach with clearly defined questions and exercises. He argued that we should ask questions about members' nearest open area, their favorite open area and look at the two in relation to where they go on holiday. His intention was to find an ideal environment against which to measure local open spaces. Neither Harrison nor Limb were very convinced by this suggestion. He argued that the senses were important: we should explore sight, sound, smell, and touch. We should find out how far people would be willing to travel to open spaces. He proposed a modelling exercise in which people could create their ideal environment through replaceable components, substituting water or a vista for a wood, for example. We should focus on the importance of water, hills, woods, birds, butterflies, and flowering plants; we should talk about individual species, using photographs perhaps; consider differences between natural and planted settings. Possibly we could have a field trip and get members to take us to their favorite sites. In effect, Goldsmith proposed to use the group as a form of group interview, similar to the practices of market researchers. Burgess, in particular was unhappy about this approach, arguing that it was much too narrowly focused and that people would not be able to talk with that kind of specificity. The directed nature of leadership would destroy any chance of exploring the ways in which people felt, thought and talked about open spaces and the natural world. We could not agree by the end of the meeting and the contest between the two kinds of group-based research prefigured many of our subsequent disagreements with the Countryside Commission.

By the meeting of January 30, 1985, we had firmed up some of our operational procedures. We agreed to pay participants £5 a session for attending, partly in thanks, partly to encourage regular attendance that was vitally important if the group was to cohere and develop a shared memory of its concerns. We agreed neighborhoods in which to convene the groups: a middle-class group from Eltham; working-class groups from Plumstead and Thamesmead; and an Asian women's group drawn from Woolwich. Burgess was now sure of the topics she wanted to use to structure the groups and persuaded the rest of the team to accept them. The sequence mirrored the phases of group development from initial cohering of members through a stage of increasing trust and intimacy to a termination phase in which members drew back into discussion of more practical issues and said goodbye to each other. The topics were: introductions; places I like; places I dislike; childhood places; management of open spaces; review and conclusions.

We discussed ways of exploring people's attitudes in a household survey. Both Harrison and Goldsmith felt that we should adhere to the terms of the original research proposal which envisaged some kind of attitude scaling technique in the questionnaire survey. The possibilities included multi-dimensional scaling; personal constructs; semantic differential scales; and adjective checklists. These would be drawn from the group discussions. Again Burgess had reservations, having used some of these techniques in her earlier work, and argued against them. Her feelings grew stronger once the group discussions were underway. By the time of the second meeting with the Countryside Commission in October 1985, in which we discussed their desire to incorporate the group material in their national household survey on recreation, Burgess was adamant that the integrity of the group discussions should not, in her terms, be "violated" by the enforcement and manipulation of statistical techniques. She felt strongly that we could not combine qualitative and quantitative research in this way. This conflict was not successfully resolved until well into the second year of the project. We were obligated to conduct some kind of social survey to meet the terms of the original proposal and, increasingly, we were being questioned about the representativeness of the feelings and values expressed in the groups. We designed a very simple questionnaire which was informed by the group discussions but we emphasized behavior rather than motivations and we used four photographs of different settings in the borough as a projective technique, rather than developing psychometric attitudinal scales (see Harrison, Burgess, and Limb 1989; Burgess, Harrison, and Limb 1988c).

Conducting the Groups

Just before the first group meeting in Plumstead, Burgess had lunch with Bramley to discuss the project and the advice and support she gave contributed much to the success of the project. Bramley emphasized the importance of the conductor in the group, stressing that the role is different from that of therapist. The conductor needs to be charismatic, inspiring the group in terms of its function; stressing the originality and importance of members' contributions; making people feel valued and important. The groups are "task-oriented, work groups" not therapeutic groups and it is vital that the conductor always bears this in mind. The group members must also feel that they have a "task," a reason for being in the group, otherwise discussions may lack sufficient focus and people will become anxious about why they are there. Having a defined task also enables the conductor to break up discussion when it has clearly strayed from the topic or when individuals are getting into deep water.

The conductor faces several problems: encouraging cohesion among group members but, at the same time, allowing conflict and dissent through which people can express their views and feelings; shutting people up without offending them; drawing out the quiet members; deciding how to let things run and for how long. Bramley stressed that Burgess should be aware of the psychoanalytic dimensions of the group, and that she should encourage, in transactional terms, the "adult" to emerge rather than creating infantile parent/child relationships between herself and the group. Members should be on an equal footing with the conductor. She emphasized this point because there would be problems of transference: people would feel fear and anxiety, a need to perform well and would hold the "academic" in awe. Members would need permission to "witter on," as it were, and to know that this was what the conductor wanted to hear.

Above all, Bramley stressed one of the central tenets of group-analytic theory: the key process and fundamental aim of the group is free association, which she described as "the music of groups." Free association encourages feelings, stories, personal reflections and constitutes the creative life of the group. She reinforced Burgess' unhappiness with structured sessions, games, exercises and projective techniques. Overstructuring is a means of allaying the anxieties of the conductor rather than a way of assisting the development and discussions of the group. But she also recommended that members be given something to think about during the week as a way of easing the start of the subsequent session—advice that proved to be very valuable, especially in the early stages of the groups.

Running the groups was immensely challenging, rewarding and enriching. It was also the most stressful field experience that we have ever undertaken. The first session of each group was by far the most important and most difficult. Members had to feel they wanted to come back next week. Burgess followed the same procedures in each group, although it was more difficult with the Asian Women's Group because we had not met three of the four women who came to the first session and we were being even more closely vetted about who we were and what our intentions were. Burgess introduced the project and the team, then introduced herself in some detail. Limb followed. Members then introduced themselves to the rest of the group. At some point in these preliminary conversations, Burgess stressed the importance of the project and the vital contribution that the group could make to a better understanding of values for open space in the borough. She emphasized the policy dimensions without making promises that decisions would be taken on the group's recommendations. She explained her role as conductor, saying that she would not be acting as a chairperson: she would not join in much, rather she wanted to listen to the members. It was vitally important that group members felt confident and that initial anxieties were eased quickly. Burgess paid considerable attention to her own body language: sitting back comfortably in her chair and relieving her own stresses and anxieties by wiggling her toes frantically inside her shoes!

The basic skills of running groups are well known (Whitaker 1985) but we encountered some specific problems dealing with our groups. For example, the Asian women demanded a much higher level of interaction from the conductor, expressing much anxiety that they were "talking about the right things." Burgess always responded very positively and contributed personal stories from her own life to reassure people and move conversation on. There were other problems with mixes of smokers and non-smokers—so Burgess asked the group at the start of each session to agree that only one person at a time would smoke. This worked well as a basic rule and was accepted by the members. Each group developed a character of its own and demanded subtle variations in the behavior of the conductor. Burgess made one serious mistake in the second session with the Eltham group when she transgressed one of the basic rules of not talking about one group with another but as our discussion of that shows (Burgess, Limb, and Harrison 1988b), it was recoverable and never happened again. She also learned very early on in the life of the Plumstead group that it was not at all appropriate to interpret what people were saying, although the expectations of the middle-class group were rather different and those members did look for some "academic" commentary. Limb and Burgess built up good personal relations with all the group members. Although two or three individuals gave some cause for concern (and Burgess worked hard in the discussions to protect them from unnecessary or damaging personal disclosures), people all expressed pleasure and satisfaction from having participated in their groups at the end of the final meeting.

The groups ran from late February to mid-July and challenged the cohesion of the research team. Burgess and Limb developed very strong personal attachment to the groups, as well as to each other through their shared field experiences. They were excited by the vivacity of the group discussions and by the insights they were gaining into the real depth, intensity and richness of people's feelings and encounters with open land. At team meetings, they would enthuse about what individuals had said or tell anecdotes of the meetings. But Harrison and Goldsmith were both excluded from the group experiences: outsiders looking in on an amazing experience. The "secondhand" nature of their experiences presented major difficulties—not least because listening to the tape recording was a very poor indication of the quality of the discussions and the contributions were often difficult to disentangle when people talked across one another. They did not know the personalities behind the voices. The problems were alleviated to some extent by the production of accurate transcripts but then both Harrison and Goldsmith were responding primarily to a written discourse that was disjointed, often difficult to follow, presented linearly and without any real indication of the moods within which things were said. There was also an enormous pile of material to read through:

each transcript runs to a minimum of 1500 lines. Harrison was able to appreciate the quality of the discussions despite these difficulties, but Goldsmith found them so far from his immediate research experience that he felt there was virtually nothing of use in the transcripts.

THE DEMANDS OF THE DATA

The research team was, quite frankly, appalled at the amount of data generated by the groups. We had hoped that an audio-typist might produce the transcripts but soon realized that no one could be expected to recognize the different voices and distentangle the arguments and discussions. Limb, therefore, undertook an initial transcription of the Eltham, Thamesmead, and Plumstead tapes while Burgess decoded the Asian women's discussions. Then we double-checked each other's work, listening through the tapes again, clarifying problems and inserting other pieces of information important for the discourse: pauses, laughter, trailing off, and interruptions. We began transcribing onto a BBC micro but the files proved too large and we transferred the material onto the departmental VAX computer. Each group session was given a unique code and the discourse was numbered sequentially line by line (see Burgess, Limb, and Harrison 1988a for full discussion). A complete set of transcripts was produced by the end of September—much later than we had anticipated.

We had not thought through our interpretive strategies in any detail before we began the groups and were now confronted with a major problem. We had made a point of writing up notes after our team meetings and that practice had been carried over into the group discussions. Immediately after the end of each group, Burgess and Limb would de-brief, discussing what the group had raised, talking about the feelings of individual members in the session and comparing equivalent sessions in other groups. These notes were then typed up and circulated to the rest of the team. At the same time, we kept case notes on all the individual members, adding snippets gleaned during coffee before the meetings and any other relevant information. These, too, were circulated to Harrison and Goldsmith. But where to go from here? We had made a commitment to the members of the four groups that we would write and send reports of their discussion, drawing out the major views and issues that had emerged over the six weeks. Thus our first priority was to read through the transcripts of each group and distill its major themes and preoccupations. These reports were written by Limb and Burgess, and then discussed with Harrison and Goldsmith before they were finalized and sent to the participants. Although never published, they proved to be of importance in our relationship with the Countryside Commission (see below).

The team decided that we would each read through the transcripts and produce an index of the material discussed. These would be combined into a systematic index of major and minor topics and areas of conversation. The indices could have formed the beginnings of a content analysis of the material, but the three women researchers felt strongly that such an approach would not do justice to the quality of the discussions in which the context of items under discussion is as significant as the items themselves. Again, Burgess and Goldsmith had different points of view. One of the central tenets of group-analytic theory is that all discourse combines both manifest and latent meanings: the surface topic under discussion is underpinned by unconscious desires and fantasies, projections and transferences which reflect deeper levels of meaning. Burgess undertook a separate interpretation of the transcripts highlighting sequences where she believed that unconscious processes were operating most clearly. With the Asian women's group, for example, there had been stormy exchanges between Rupa and Surjit which had resulted from a mix-up over a Bank Holiday. The group met for its third meeting after a week's break. The offended party did not arrive for the start of the meeting and the early discussions were full of symbols of death: the dangers of snakes at home in India; fearful encounters with strange animals; and cemeteries. When Surjit did finally arrive, she said nothing for some while until she suddenly burst in with a life-enhancing story of the birth of her first child. The group was "saved." Goldsmith was not able to accept the validity of this kind of interpretation and felt that it offered no help in meeting the project objectives.

Even now, we find it difficut to handle these interpretations in a disciplinary context. Clearly the manifest content of the discussions is of immediate relevance and interest to our various academic and policy communities but theoretically, latent content is of great importance. We have confined ourselves to offering a group-analytic interpretation of the discussions of the Eltham group as a way of highlighting the vital importance of approaching discourse with an appreciation of its complex, contextual nature (Burgess, Limb, and Harrison 1988b) but we do not propose to undertake any further publications which focus so closely on the psychoanalytic dimensions of the work. Not only is it beyond our competence, it is also a gross intrusion of our members' privacy and their understanding of the purpose of the project. To protect individuals and to maintain the work orientation of the groups, Burgess did not encourage personal revelations that were not relevant to the topic. But the issue highlights major problems of confidentiality and we have, in one respect, breached our commitment. We assured participants of the confidentiality of the discussions and promised to use pseudonyms in any publications. This we have done. But our major sponsor required that copies of the research data be lodged with the ESRC data archive and we, too, want other researchers to have access to the material. So we have removed personal details from the discussions that might enable people to be identified and requested that researchers obtain permission from us before they can use the material.

As a way of trying to find out what people were thinking about, and what aspects of ecology they were particularly interested in, Goldsmith undertook an ecological content analysis of the four transcripts, endeavoring to relate his own interests and research questions to the data. Not only was he disappointed at what he interpreted as the lack of any ecological awareness among the group members, he expressed some frustration that Burgess had not pursued matters of ecological interest when they did arise spontaneously in the discussions. For example, in the Eltham group, Elaine had made passing reference to practices of coppicing. "But" Goldsmith said mournfully at a team meeting, "you didn't ask her what she knew about it." He wrote a brief paper on ecological awareness suggested by his content analysis that Harrison and Limb read. The two women felt that the ecologically-based analysis did not really do justice to the group discussions where ecological matters were expressed in terms of people's daily lives. At this stage, Harrison, Burgess, and Limb took the major role in interpretation and publication of results from the project.

Our interpretive strategies are emerging as we work with the transcripts. We find the most effective means of presenting material from the groups is to construct narratives around particular themes. We trace issues through and between groups' discussions and present them in terms of what seem to us collectively, to be the most significant themes through which people express their values and feelings. But it is very hard to report the verbatim material in a way which maintains the integrity of the group discussion. It takes up considerable space in publications with tight word limits, for example. The papers we have written so far reflect our major preoccupations as a cross-disciplinary research team and so countryside recreation (Harrison, Limb, and Burgess 1986a, 1986b); values for nature in the city (Harrison, Limb, and Burgess 1987); two methods papers (Burgess, Limb, and Harrison 1988a, 1988b); and a paper on the meanings of open land (Burgess, Harrison, and Limb 1988c) have come first.

We want to work in much greater detail with the transcripts but are not yet sure about how best to proceed. We might profitably turn to cultural studies and linguistic theory in order to progress beyond the narrative stage of our work. We have come to believe that environmental values are socially and culturally constructed, and that it is possible to explore the semantic structures within which people represent their experiences and within which meaning is signified. There seem to be a number of key dialectics through which environments are valued and convey social meaning, not the least powerful of which is *country* and *city*. We would like to pursue the relationship between cultural myths of country and city and people's situated, practical realities. We would like to explore the hegemony in landscape values and see the extent to which our groups accept or contest dominant heritage, conservation, and property values. But first, we feel that we must address the policy issues relating to the research, especially in the context of Greenwich borough and open-space planning in London. That was a major part of our original commitment to our group members and for reasons outlined next, we have not yet been able to deliver our promise.

CONTACT WITH THE POLICY FORMING COMMUNITY

Our contacts with the staff of the Countryside Commission have included personal meetings with officers of the Commission and attendance at two seminars convened by the Commission as part of their recreation policy review. Indirect contacts include publications sent directly to them; presentations at national conferences at which officers were present; and contributions to their Rangers' Training program. In their separate ways, each of these contacts had a significant bearing on our research but two aspects are most important. First, the Commission's reaction to the nature of qualitative research as a basis for informing policy and second, their use of our work to characterize what the general public want of the countryside.

As an agency that had commissioned several commercial market research companies to undertake research for them, the Commission was familar with conventional methods of social survey whereby once-only small group discussions are used as precursors to questionnaire surveys (see, for example, the Access Study, Countryside Commission 1986b). We had intended to use discussion groups in this way; however, the richness and complexity of the material generated by the in-depth discussion groups, presented real methodological problems for us as researchers and for the Commission as clients. The conversations in the groups provided the Commission with numerous examples of what they had been hoping to hear—discussion about why people value the countryside, what kind of countryside is appealing, why people decide to visit it and what aspects of their visit are enjoyable or unpleasant. Listening to extracts of the early discussions, Commission staff were enthused by the quality of the material and encouraged by the overall approach—so much so that they wanted to sit in on a group! Burgess refused. They also responded favorably to the Group Reports that were sent to them in the late summer of 1985. These reports were prepared for the group members and were written as a narrative account of the major themes that emerged over the life of each group. In each report, topics were woven around extended quotations taken from the transcripts. Our liaison officer in the Commission said these reports came as "a breath of fresh air" because they were written in straightforward language; they conveyed an authenticity that was totally lacking from other research; they highlighted a diversity of material; and perhaps, above all, because they revealed the enormous value of the countryside for people from all walks of life. We attribute much of the support the Commission subsequently gave to qualitative work in their review of recreation policies to the impact of the Group reports.

Contesting the Credibility of the Methodology

Notwithstanding the warm reception the reports received, on their own they did not provide the Commission with a systematic analysis of environmental attitudes and values, recreational motivations and benefits, or any indication of how the findings from the group discussions might relate to recreation policy. In pursuing this latter objective, officers of the Commission requested a meeting in the autumn of 1985 to discuss how our detailed analysis of the transcripts might proceed. The meeting proved to be a turning point in our work. At this stage, we were trying to develop a method of interpretation that would be systematic and reflexive and faithful to the sentiments of the group members. Although the groups had been convened to explore people's feelings for open land, in practice, members found it impossible to talk about open space without also talking about how they lived out their everyday lives. Moreover, it became clear over the extended life of the groups that seemingly neutral terms such as countryside and countryside visits had different meanings for different groups of people. Explaining how people formed an attachment to countryside thus involved complex interpretations of their structures of feeling in which aspects of culture, society and experience were intimately bound together. We needed time to develop and implement a set of ground rules for interpreting the transcripts and to refine these interpretations in the light of other work. Only then, we felt, could we move on to explore policy implications of our findings. As policy formers, the Commission were much more concerned to use the findings of the group discussions directly. They wished to incorporate these in the review of countryside recreation policy and specifically, in a national household questionnare due to be undertaken as part of this review.

When we met to discuss our strategy for interpreting the transcripts, these policy considerations assumed overriding importance. The officers were much less interested in what they saw as the "academic" and theoretical aspects of the work. Working to a timetable dictated by the policy review, they were not to be deflected by our assertion that progress on providing a fuller interpretation of the transcripts would take time. Could we not provide them with a "pale shadow for policy?" In this seemingly simple request, the officers acknowledged the insights, richness and essential worth of the groups, and their own frustrations and uncertainties about how qualitative material could be used to inform policy.

Our agreement to prepare a report that would focus on popular values for the countryside together with some indication of how these attitudes impinge on countryside recreation policy, deflected us from our original intention which had been to prepare a report on the role of open spaces in the urban area. In turn, it provided an opportunity for us to contribute to an active policy debate before firm policies had been formulated and conventional processes of public consultation set in motion. It was an opportunity that we had not foreseen at the outset of our work and one from which both we and the Commission benefited, as they acknowledge in their review of the impact of research in countryside recreation policy development (Phillips and Ashcroft 1987).

Our unwillingness to agree to formulate a series of attitudinal statements suitable for incorporating into a questionnaire survey (which had fuelled much of our internal disagreement) reflected our growing conviction that approaches based on psychometric techniques are poor indicators of environmental attitudes, values, and meanings. All the evidence in the group discussions pointed to the importance of deeply embedded cultural values in moulding attitudes to countryside, as well as to the importance of significant events in people's life histories. Psychometric techniques assume that feelings and attitudes can be expressed in isolation, and that these complex associations of feeling can be dissected, reduced to a series of pre-selected categories and measured. Through the in-depth groups, we find that what is important is the holistic experience of countryside—the company of others, the intangible properties of the natural world, the life context, and the symbolic meaning of the countryside as a better way of life. Responses to psychometric and projective techniques used in questionnaire surveys and single group interviews can of themselves only ever be a pale shadow of this holistic experience. We failed to convince the Commission of the shortcomings of these kinds of techniques, committed as they were to the research. However, they were sufficiently encouraged by the preliminary findings from the group discussions to commission a commercial agency—the Qualitative Consultancy—to undertake a series of once-only group meetings in other parts of the country as a precursor to the questionnaire survey (Phillips and Ashcroft 1987).

We sent our report, *Popular Values for the Countryside,* to the Commission in January 1986. It outlined the methodology we had adopted in the research, including the principles of Group Analysis and pointed out the ways in which both the theory and practice of group-analytically orientated work groups differ from conventional single group interviews. The central part of the report was devoted to a systematic interpretation of popular values for the countryside. The final section identified a number of points arising from the research and on countryside recreation policy. We heard nothing from the Commission until the middle of March when we were invited to attend a presentation by the Qualitative Consultancy of their preliminary findings.

This meeting, held in the Commission headquarters at Cheltenham, turned into a fracas. It was not chaired, the representatives from the Qualitative Consultancy did not know of our work, no attempt was made to address the findings of the two research contracts or to examine what weight might be given to them in the development of the Commission's countryside recreation policies. From our point of view, the theoretical and methodological differences between once-only groups and our own in-depth groups were so great that the meeting became something of a contest between competing methodologies. To others present, it appeared to be simply an "academic squabble." We came away convinced of the Commission's ignorance about the different methodologies and the implications of these differences for the findings and usefulness of qualitative research. We departed, fuelled with a determination to expound these differences through presentations to a wider public arena. This we did. First in an article written for *Landscape Research* (Harrison, Limb, and Burgess 1986b), and second in two methodological articles published in *Environment and Planning A* (Burgess, Limb, and Harrison 1988a, 1988b).

Acquiring Credibility through Peer Group and Public Review

We presented the burden of the first article to the Landscape Research Group conference in early May 1986, in a session chaired, coincidentally, by the Commission's Assistant Director (Policy) who had been present at the Cheltenham meeting. Later that same month, his office sent us an invitation—alongside other "normal consultants"—to comment on the discussion paper on future recreation policies (Countryside Commission 1986a). In June, we were invited, together with local authority officers and voluntary groups, to attend one of the regional seminars held as part of the national consultation exercise. The credence given to our views in this early phase of public consultation was unexpected in the light of our previous encounter. In these formal discussions, we were called upon to present the views of ordinary people and not just the views of interested academics or researchers. In effect, on these and other occasions when invited to contribute to training courses for countryside rangers, we have been asked to act as the voice of the public. We see dangers in being identified in this way, but it does suggest that our work has gained credence precisely because we bothered to find out what people really want of the countryside. The favorable reactions to our work among rangers and other professionals who have contact with people at the grass roots level, has given us a great deal of pleasure and has allowed us to make contributions to an area of recreation policy which we never anticipated.

Our written response to *Recreation 2000* was sent in August and was formally acknowledged. But later, in October 1986, we received a letter from one of the officers who had acted as our contact during the research project. In it, he expressed pleasure at seeing our paper in *Landscape Research* and said that he had decided to use it as our response to the discussion paper to aid their deliberations. Clearly at this point, being able to refer colleagues to a published paper was more helpful than either our formal response or our specially prepared report per se. To quote: "Your research offers us a number of ways of resolving some long running and up until now rather sterile debates about the role of recreation sites and the legitimacy of promoting countryside and to whom." The decision to publish our work—and our ability to do so without any constraints being placed upon us by the research sponsors—had been vindicated.

We need to add a postscript to the story. As we were finishing this chapter, Harrison attended a day conference on *Public Perceptions of the Countryside.* The papers were distributed, bound together in book form, at the end of the meeting (Miller and Tranter 1988). The collection contains a paper by Sir Derek Barber, the Chairman of the Countryside Commission, which expresses the Commission's views about people's perceptions of the countryside. We were surprised and delighted to find that the Commission has accepted fully the value of qualitative field research:

> Most research techniques ... require people to react instantly to a series of questions put by strangers either on the doorstep or the telephone. This must surely have its limitations in measuring what people feel.... An opinion about a straightforward concept is all one can expect to obtain—with the strong risk that what really matters remains hidden. Getting at how people form their views and how they are influenced requires a much more subtle approach which is found in a method called 'Qualitative Research.' (Barber, 1988, p. 45).

Sir Derek then defines qualitative research as working with groups. We were also surprised to find that a major part of his paper that presents the substantive material on public "perceptions" is taken directly from our report on *Popular Values for the Countryside.* We are gratified that the research should be presented as the Commission's views but, predictably perhaps, neither we nor the ESRC as our major sponsors, are acknowledged in the paper. As one of our colleagues remarked, it normally takes six or seven years for research to affect policy in any material way. We should be content that we have apparently managed it in two!

CONCLUSIONS

Given this opportunity to reflect on the last three hectic years, we have become much more conscious of the ways in which our internal group dynamics and the policy demands of our minor sponsor have influenced the progress of the research. Our conclusions about the worth of the project differ. We all accept that there remains the problem of crystallizing "popular values for the countryside" and we accept also that we never resolved the tensions between qualitative and quantitative social science research. Over the life of the project, our theoretical and methodological conflicts were only "resolved" for pragmatic reasons—in order to satisfy our contractual obligations and not because we had developed an integrated research strategy. There remain profound differences between that part of our study designed to research people's behavior and that greater part concerned with the social meanings and values of open space.

In the light of our collective experiences, Goldsmith believes that we have learned what a few people cherish and, to a limited extent, what motivates people to visit the countryside or open space. Our research has suggested a possible methodology but still lacks the means of determining the views of a wider, more representative audience. He believes that because the groups were not recruited in a definable way, the findings remain unrepresentative of the wider population and, therefore, are of little value for countryside management agencies such as the Countryside Commission or the National Trust. To this end, he believes that the discussion groups need to be recruited more systematically and their subject matter requires a clearer structure. While the group-analytic approach investigates people's attitudes in depth, he believes the method to be time-consuming and expensive and ultimately of only limited value to policymakers.

From the two geographers' perspective, it is important to understand the vital links between nature and culture for it is these links that ground the activities of management agencies. Geographers share a way of seeing that enables them to move reasonably comfortably between Group Analysis, values and policy and as a biogeographer, Harrison fulfills a useful role in disseminating the results of the project to ecologists and conservationists. On a methodological level both Burgess and Harrison are convinced of the worth of in-depth groups as a freestanding method of enquiry. The new theoretical horizons that have been opened up through the group discussions have emerged in consequence of the theoretical strength of Group Analysis and its clear guidelines for the professional conduct of task-orientated groups. The work has allowed us all to explore new questions that other approaches do not raise and Burgess and Harrison are optimistic that they will be able to forge new explanations for environmental values that are grounded in the shared meanings of culture and the everyday realities of modern life. On an applied level too, the disciplinary and ethical demands of conducting qualitative research in a professional manner and of carrying this work into the policy-making community have only served to reinforce a commitment to enabling people to speak for themselves in public arenas where their voices are seldom heard. We are all equally convinced of the need for independent academic researchers.

NOTE AND ACKNOWLEDGMENT

1. We wish to acknowledge the financial support of the Economic and Social Research Council (contract no. D0023217) and the Countryside Commission. The transcripts of the group discussions, together with the survey results, are lodged with the ESRC Data Archive at Essex University.

REFERENCES

Abercrombie, J.

1983 "The Application of Some Principles of Group-analytic Psychotherapy to Higher Education." Pp. 3-16 in M. Pines (ed.) *The Evolution of Group Analysis.* London: Routledge and Kegan Paul.

Barber, D.

1988 "Perception of the Countryside: The Views of the Countryside Commission." Pp. 44-48 in F.A. Miller and R.B. Tranter (eds.) *Public Perceptions of the Countryside* Reading: Centre for Agricultural Strategy, Paper 18.

Bramley, W.

1979 *Group Tutoring: Concepts and Case Studies.* London: Kogan Page.

Bramley, W. and Wood, P.

1982 "Collaboration, Consultation and Conflict." *Journal of Geography in Higher Education* 6:5-20.

Burgess, J.

1978 *Image and Identity: A Study of Urban and Regional Perception with Reference to Kingston upon Hull.* Hull: Occasional Paper 23, University of Hull.

1982 "Filming the Fens: A Visual Interpretation of Regional Character." Pp. 41-65 in J. Burgess and J.R. Gold (eds.), *Valued Environments.* London: George Allen and Unwin.

1985 "News from Nowhere: The Press, the Riots and the Myth of the Inner City." Pp. 180-220 in J. Burgess and J.R. Gold (eds.), *Geography, The Media and Popular Culture.* London: Croom Helm.

1986 "Crossing Boundaries: A Group-analytic Perspective in Geographical Research." *Group Analysis* 19:235-243.

Burgess, J., Limb, M., and Harrison, C.M.

1988a "Exploring Environmental Values through the Medium of Small Groups. Part One: Theory and Practice." *Environment and Planning A* 20:309-326.

1988b "Exploring Environmental Values through the Medium of Small Groups. Part Two: Illustrations of a Group at Work." *Environment and Planning A* 20:457-476.

Burgess, J., Harrison, C.M., and Limb, M.

1988c "People, Parks and the Urban Green: A Study of Popular Meanings and Values for Open Spaces in the City" *Urban Studies* 25:455-476.

Burgess, J. and Wood, P.A.

1988 "Decoding Docklands: Place Advertising and the Small Firm." Pp. 180-220 in J. Eyles and D. Smith (eds.), *Qualitative Methods in Human Geography.* Oxford: Polity Press.

Cosgrove, D.

1984 *Social Formation and Symbolic Landscape.* London: Croom Helm.

Countryside Commission.

1986a *Recreation 2000. A Discussion Paper on Future Recreation Policies.* Cheltenham: Countryside Commission.

1986b *Access to the Countryside for Recreation and Sport.* Cheltenham and London: Countryside Commission and Sports Council. CCP 217.

Eyles, J. and Smith, D. (eds.).

1988 *Qualitative Methods in Human Geography.* Oxford: Polity Press.

Foulkes, S.H.

1975 *Group-Analytic Psychotherapy.* London: Gordon and Breach.

Goldsmith, F.B.

1983 "Evaluating Nature." Pp. 233-246 in A. Warren and F.B. Goldsmith (eds.), *Conservation in Perspective.* Chichester: Wiley.

1990 "The Selection of Protected Areas." In I. Spellerberg, F.B. Goldsmith, and M.G. Morris (eds.), *Scientific Management of Temperate Communities for Conservation,* British Ecological Symposium *31.* Oxford: Blackwells.

Green, B.
1981 *Countryside Conservation: The Protection and Management of Amenity Ecosystems.* London: George Allen and Unwin.

Harrison, C.M.
1983 "Countryside Recreation and London's Urban Fringe." *Trans. Inst. Br. Geog.* 8 (New Series): 295-313.

Harrison, C.M., Burgess, J., and Limb, M.
1986 *Popular Values for the Countryside.* Unpublished report prepared for the Countryside Commission.
1989 "Popular Values for the Countryside." Pp. 43-57 in B.J.H. Brown (ed.), *Leisure and the Environment.* Proceedings of Conference of Leisure Studies Association, No. 31, April 1987. London: L.S.A.

Harrison, C.M., Limb, M.L., and Burgess, J.
1986 "Recreation 2000: Views of the Country from the City." *Landscape Research* 11: 19-24.
1987 "Nature in the City—Popular Values for a Living World." *Journal of Environmental Management* 25:347-362.

Jackson, J.B.
1984 *Discovering the Vernacular Landscape.* New Haven: Yale University Press.

Little, C.E.
1975 "Preservation Policy and Personal Perception: A 200-million-acre Misunderstanding." Pp. 46-58 in E.H. Zube, R. Brush, and J.G. Fabos (eds.), *Landscape Assessment: Values, Perceptions and Resources.* Stroudsburg: Dowden, Hutchinson and Ross.

Mabey, R.
1980 *The Common Ground: A Place for Nature in Britain's Future?* London: Hutchinson.

Meinig, D. (ed.).
1979 *The Interpretation of Ordinary Landscapes.* Oxford: Oxford University Press.

Miller, F.A. and Tranter, R.B. (eds.).
1988 *Public Perceptions of the Countryside.* Reading: Centre for Agricultural Strategy, Paper 18, University of Reading.

Mostyn, B.J.
1979 *Personal Benefits and Satisfactions Derived from Participation in Urban Wildlife Projects.* Shrewsbury: Nature Conservancy Council, Interpretative Branch.

Penning-Roswell, E.C. and Lowenthal, D. (eds.).
1986 *Landscape Meanings and Values.* London: Allen and Unwin.

Phillips, A. and Ashcroft, P.
1987 "The Impact of Research in Countryside Recreation Policy Development." *Leisure Studies* 6:315-328.

Pines, M. (ed.).
1983 *The Evolution of Group Analysis.* London: Routledge and Kegan Paul.

Qualitative Consultancy.
1986 *Qualitative Research to Explore Motivations behind Visiting the Countryside.* (Mimeo, The Qualitative Consultancy, London.)

Shoard, M.
1980 *The Theft of the Countryside.* London: Temple Smith.

Usher, M.B. (ed.).
1986 *Wildlife Conservation Evaluation.* London: Chapman and Hall.

Warren, A. and Harrison, C.M.
1984 "People and the Ecosystem: Biogeography as a Study of Ecology and Culture." *Geoforum* 15:365-381.
Whitaker, D.S.
1985 *Using Groups to Help People.* London: Routledge and Kegan Paul.

SPLITTING IMAGE: "PURE" AND "APPLIED" RESEARCH IN THE CULTURE OF SOCIOLOGY

Alan Prout

INTRODUCTION

It is both a satisfaction and a frustration of qualitative sociological research that a line can be drawn under the account of a particular episode of fieldwork only with great difficulty. However much one tries tidily to pack away a research project and get on with something new, fresh insights into old experiences continue to occur, appearing on the one hand as frustrating sidetracks to new business at hand and on the other creating a satisfying sense of organic wholeness and continuity. Such is especially the case with the fresh perspective which distance can allow on "completed" research. This is not only a function of passing time but also of a new viewpoint from which old research can be seen, as current concerns shed a different light on previous experience. In this chapter I am going to set out some reflections on the context and process

Studies in Qualitative Methodology, Volume 2, pages 169-188.

ISBN: 1-55938-023-3

of two pieces of qualitative research both of which have formally "ended" but about which I continue to think. It is no accident that the focus of these reflections is the relationship between qualitative research and practice; that is to say between research and its implications for the practical work of those like teachers and health professionals who are so often its objects/subjects (in itself a revealing ambiguity). My concern for this relationship reflects the shift in my own role: the fieldwork projects I will refer to were carried out as a full-time researcher (in one case as a doctoral student and in the other as a contract researcher in a university) but I now teach sociology to health professionals. This shift from research to teaching has, I have found, not only raised questions about the salience of sociology to the practice of the health workers whom I teach but has also caused me to reconsider some of my previous views and experiences.

In true qualitative manner my discussion is based on reflection about two instances from my own research biography, although I hope that through these I can raise issues of more general significance. In particular I will use my own experience of these two research projects to suggest that the culture within which sociological research is carried out tends to discourage and devalue thinking about its practical implications. Since I think this is unhelpful to both sociologists and the non-sociologists who might benefit from our work, I want to suggest that we have a responsiblity to examine and, if necessary, reform our own practices.

The circumstances in which this chapter first came into being illustrate the hesitation, even distaste, with which sociologists greet the notion that their work might have practical implications. It began when I was asked to contribute a seminar paper on the policy implications of my Ph.D. research.[1] My first reaction was that there were no such implications, a response which the seminar organizer told me he had found to be a common one. So interesting did this seem to be that I offered a paper exploring the possible social meaning of my own apparently typical reaction. A discussion of this topic forms the first part of the chapter, where I argue that my reaction articulates some features of the culture of recent British sociology. The second part of the paper widens into a discussion of how sociologists might rethink our own practices in relation to the practical implications of our research.

BACKGROUND TO TWO RESEARCH PROJECTS

Before starting these discussions in detail it is necessary to fill in some of the background to the two research projects with which I was involved between 1982-86. The first was a piece of contract research, that is, it was done in a university department on a temporary contract funded externally. The second was my doctoral research, which I completed as a part-time student. The tensions between sociological work carried out in these different contexts will be a major feature of my analysis, exemplifying two contrasting stances toward practical implications and highlighting some aspects of the culture within which sociological practice itself takes place.

My contract research was concerned with parenthood education in secondary schools and was done jointly with Shirley Prendergast at the Child Care and Development Group in Cambridge. It was funded by the then Health Education Council.[2] The origins of the project were in earlier work that we had done on children's health knowledge (Davies, Prendergast, and Prout 1978; Prendergast and Prout 1980; Prout 1985b). As part of this earlier study we had interviewed teenage boys and girls about their understandings of a variety of "health issues" including motherhood and childcare. This material had suggested that teenagers were far from being dominated by the romantic, glamorized and idealized views of motherhood that the literature suggested. They seemed on the contrary to be acutely aware of the problems, difficulties and "negative" aspects of full-time child care but this personal knowledge, gathered from kin and neighbors or from direct experiences of childcare, was difficult for them to reconcile with the normative rhetoric of maternity found in public discourse (Prendergast and Prout 1980; Prout 1984).

This work was based entirely on one-to-one interviews with teenagers and partly as a result of the way the data was collected was somewhat passive and static in conception. In particular it had little to say about how school children might *use* their understanding of parenthood in a specific setting. On the other hand the perspective of the (by then not so) "new" sociology of education (for example, Young 1971; Gorbutt et al. 1972) suggested that the social construction and control of knowledge in the classroom is central to the schooling process and some preliminary classroom observation had indeed confirmed that it would be fruitful to examine in detail the interaction between the knowledges about parenthood that pupils brought to school with them, and of which our interview material seemed to have given us a glimpse, and those knowledges that seemed to form the basis and assumptions of "parenthood education" in school, which was at the time a growth area in the secondary curriculum. Our research proposed to follow up systematically this preliminary observation with a longitudinal study of groups of pupils as they passed through different forms of parenthood education. The Health Education Council eventually funded a version of this proposal.

From its point of view the HEC seemed to have two rather different interests in the work. The first arose from the HEC's role as a curriculum development agency. In the previous decade it had sponsored the production of teaching and learning materials for schools' health education, but had funded little research on how this was being used at the classroom level. Our research promised to give a picture which, while based on only one particular topic area, might suggest more general issues and problems for the agenda of educational policymakers in the field of health education. The second HEC interest was much more derived from the medical mainstream of its work. At the time of our application for funds a major health education campaign on infant mortality was being planned. One aspect of this concerned the school curriculum as a medium for preventive education, especially the encouragement to early use of antenatal care during pregnancy.[3] Our research promised to guide the implementation of these plans at the classroom level.

After a period of negotiation and modification of our proposal, we were funded to carry out an intensive study of the process of parenthood education in a small number of secondary schools, using methods with a large qualitative component. In the event we followed four classes of pupils through different types of parenthood education, using a formal interview before and after their course and observing the classes they attended, in two cases over an entire school year. This material, together with that gathered in informal interviews, allowed us to build up a picture of the social construction of knowledge in the classroom, tracking the knowledge and learning of the pupils and tracing the interaction of knowledges from within and without the classroom (see Prendergast and Prout 1987).

When we began the project we were aware that parenthood education was a potentially ideologically contentious one. In fact the intensity of political interest in the area increased as our work progressed, partly as a result of Conservative government initiatives on the curriculum and the pressure on them from the "new moral right" on related aspects of schooling policy, especially sex education. Toward the end of the project we applied for an additional period of funding in order to disseminate the findings of the research. In the new climate that had developed this had become somewhat controversial, for reasons discussed below, but after a battle we did eventually succeed.

My doctoral research also had its origins in the original study of children's health knowledge mentioned above, but in this case the starting point was in some work we did on children's notions of infection, resistance and cure (Prout 1985a). While analyzing interview material on this topic I became interested in the way in which children contexted their accounts of everyday ailments such as colds by reference to negotiations with their mothers about what was serious enough to warrant time off school. It was also clear that in contrast to the active role which the children presented themselves as playing in the construction of their sickness episodes, the relevant medical sociological literature tended to cast children as passive.[4] Furthermore it tended to ignore the possible relationships between children's sickness and other aspects of childhood as a social and cultural construction: schooling, play, friendship, subcultures, and so forth (see Prout 1987a). I became interested in pursuing these issues as the basis of doctoral research and registered as a part-time student at the Centre for Medical Anthropology, Keele.

The outcome of my work was an ethnographic study of sickness absence based on participant observation among a class of fourth year primary school children during the period just before they transferred to secondary school at the age of eleven. My analysis related variations in the rate of absence to phases in the trajectory of transition to secondary school, particularly those that turned around a series of grading tests that the children took during the middle of the last term at primary school. Within this sickness was analyzed as a dominant symbol (Turner and Turner 1978, appendix) of the process of transition to secondary school. This period was seen as having liminal-like features of a rite of passage, such that the cultural performance of sickness by children, parents and teachers expressed both the dominant normative rules of work discipline, leisure and gender and the normative conduct that legitimates the illegitimate uses of "going sick" (Prout 1987a).

PURE AND APPLIED: A SPLITTING IMAGE

These two pieces of work were carried out over the same period of time, although the actual periods of formal data collection did not overlap. I have described them in some detail partly to provide a context within which the reader might understand the discussion that follows. The account also underlines the extent to which there were similarities in the underlying approach: both used fieldwork methods in a school setting; both followed schoolchildren over a period of time; both involved extensive interviewing with children; both had their origins in a concern with children's cultural knowledge (of parenthood in one case and sickness in the other); and, most important for the discussion in this chapter, both were rooted in theoretical considerations about how that knowledge is acquired and used by children. Although the theoretical frameworks of the studies developed in quite different directions, it was not the case that one (the applied work on parenthood) was atheoretical while the other was theoretically informed.

It was not, however, until I was asked to speak to the policy implications of my doctoral work that I realized how differently I had come to think about them: despite their underlying consonance I had come to see the work on parenthood education as self-evidently "applied" but my doctoral work as "pure." As I further reflected upon this I realized that what I was doing was not so much *denying* as *resisting* the suggestion that my doctoral work might have policy implications. It was as if the very thought that this particular piece of work might have policy implications was in some way offensive, although at a rational level it was undeniable that there were such implications to it. Further reflection suggested that what I had done (and what the seminar invitation had helped to raise in my consciousness) was to split my involvement and appreciation of the two projects into a system of difference. Their relationship within my mental set was not based on their many similarities and overlaps, but on their difference and separateness. My Ph.D. research had to come to stand for "proper" sociology, my contract research something different, something less, and something not as pure. There seem to be four possible explanations for how this splitting may have occurred and I will consider each in turn. The first works at an individual level. Splitting may be a coping mechanism for dealing with work overload, a means of focusing and summoning up the energy needed to sustain both doctoral and contract work. However, while this accounts for the compartmentalization in itself, it does not explain how one side of the system of difference came also to be associated with something more worthy and sociologically pure.

A second possibility lies in my own career and biography. I graduated in sociology in the early seventies at a time when sociology became identified with a radical political stance, a turning away from "useful" studies into theory and social criticism. In the period before this, sociology had split itself from social administration and turned away from empirical study more generally. By the time empirical work found favor again it had lost its connection with policy and practice concerns, but the image of policy research as tainted by atheoreticism, reformism, and problem-solving rather than problematizing, lingers with me. In fact in our work on parenthood education we had attempted to avoid vulgar empiricism and had taken a critical view of its political assumptions, but despite having this direct experience of applied work that at least attempted to avoid some of the pitfalls, I seemed to have reproduced the split at another level of consciousness.

A third strand of explanation might be found in the perhaps inevitable negotiations, compromises and conflicts of funded research. In our parenthood education study we certainly experienced these. Our own preference, for example, would have been for an entirely enthongraphic study, but we knew that this was unlikely to be funded and indeed "What are you measuring and how?" were questions on the lips of the HEC administrators with whom we discussed our proposal before it was finally submitted. These administrators were, incidentally, not unsympathetic to qualitative research, but they did understand the politics of research funding better than we did and we were glad to follow their advice, so long as it did not erode the core of the study we wished to do.

As our work progressed it remained the subject to political attention, especially that arising from battles over the health education curriculum. For many on the political right health education was synonymous with sex education and the government, influenced by lobbies in and around the Conservative Party, began to appoint HEC Council members who seemed to bring a brief to curb the HEC's role in this field.[5] The most notorious example was the withdrawal of a resources list on sex education written for teachers but which contained items, notably a book by Jane Cousins called *Make It*

Happy, which outraged certain sections of moral opinion. Certainly, both of the DES nominees to the HEC at this time were active members of the Responsible Society.[6] Our work became caught up in these conflicts because it was perceived as not supporting a morally acceptable view of the family. This was not unexpected and in a sense we were fair game since we did in fact take the view that a single and simple view of "the family" was an unrealistic basis for the curriculum. Our research supported this by portraying the variety and complexity of teenagers' understandings of parenthood. In any case our work also dealt with pupils as active subjects in the classroom process, rather than passive absorbers of curriculum content. This directly undercut the transmission view of the curriculum implicit in the notion that it should only be concerned to convey supposedly dominant moral values to children.

Looking back I do not object to our research being treated to a political attack—that is in the nature of doing sociology. What made the experience an unpleasant one was that the politics were not openly discussed but formed a half-hidden agenda to the decision-making process by which it was decided to fund the dissemination of our work. It was, therefore, very difficult to tackle the political underpinning of the discussion in any direct or rational way. This was a chastening, demoralizing and a strong contrast to the prevailing norms (if not always the practice) of academic and intellectual discourse. It is this contrast, rather than the details of the funding process, that I want to draw attention. Doctoral work does not take place free of a particular context and its constraints, but those constraints are usually strongly rooted in the traditions of academic freedom. In contract research those traditions are much weaker, more susceptible to erosion and have to be continually and actively repaired.

In addition to this, doctoral work was for me, as I suspect it was for others, the most personal of research. Perhaps (in the cottage industry of social science, at least) the individualized, often isolated and at times even cloistered context of its production result in its investment with the subjective significance and personal identity that I have found in myself and observed in others. Again I feel forced back into the language of purity, if not the sacred and profane, although at one level the doctorate as a rite of entry to the profession is deeply profane—whole livelihoods can depend upon it. Nevertheless, I developed a strong sense of protectiveness toward my Ph.D. research, not only wanting to hygienically shield it from the distasteful, and wearying, politics of research funding but also placing the two in cognitive/emotional opposition.

Finally, and as a last explanation for this splitting, I want to draw attention to a further set of problems faced by contract researchers, this time not with their funding bodies but in their relationships with permanent teaching staff and the departments within which they are based. Sue Scott (1984) has described in some detail the marginalization which she experienced in this role, although for her gender and status relations were inextricably entwined together. Gender apart, however, her experience chimes closely with my own.

The research unit was regarded as largely separate from the department's mainstream teaching and social science research work: it was also physically somewhat separate; most of its members were researchers financed by externally raised funds and doing broadly "applied" work; these contract researchers were not included on the governing committees of the department, although this has since been remedied. Overall there was a pervasive feeling among contract researchers that they were regarded as vulgar fact grubbers, and in a sense it hardly mattered whether this was an accurate perception or not. Scott (1984, p. 168) described her response to this type of situation as follows:

> (Such problems) . . . seemed strange to me at the time, made me much more defensive about the research and often caused me to be overtly critical of it myself. In an atmosphere where theoretical work was the norm, and where teaching and writing books was seen as 'proper' academic activity, where many people felt qualitative methodology to be suspect, and where the research topic was seen as inconsequential, the fact that two young women, lacking the academic respectability of publications and higher degrees, were not treated as equal is no longer surprising.

I feel sure that this sort of experience is widespread, indeed it is a commonplace of conference tea-room chat among researchers. In this sense the split which I experienced/constructed between pure and applied work has a contemporary as well as a historical counterpart in the character, culture and organization of British sociology.

RELATING SOCIOLOGY TO PRACTICE

I am no longer a contract researcher or a Ph.D. student. Nevertheless, in my new role as a teacher of sociology to health workers I find that the tensions and splits I have described above continue to make themselves felt. I must engage with and make a contribution to their education and training that they and I feel is worthwhile but not uncritical.

One frequently used framework for thinking about this issue is the debate about the relationship between research and policy. The earliest conceptualization of this relationship was the so-called "engineering model." In this research is seen as having a direct input into policy decisions which, it is claimed, are thereby guided in a rational manner. It is now recognized that this is neither an adequate description of the policy-making process nor an appropriate prescription of the political role of social researchers (see for example, Bulmer 1982). Policy changes are in practice rarely directly influenced or initiated by research findings, which tend to feed indirectly into policy debate, rather than directly into a particular policy decision. These points form the basis of an alternative approach, sometimes called the "enlightenment model" of the research-policy relationship. In this social researchers are seen as having a real but diffuse contribution to policy-making. In particular social research can illuminate policy debates without necessarily recommending particular courses of action and it can help set the agenda of policy debates by drawing attention to new issues and areas of concern. The enlightenment approach was not developed as having specific relevance for qualitative methods but it seems that such research, because it offers interpretation and illumination rather than claims to the discovery of objective facts, does have a particular affinity with it.

Finch (1986), while accepting it as an improvement on the engineering approach, goes on to suggest some weaknesses with the enlightenment model. She suggests that both can be elitist. In both models research is addressed to the powerful, revealing to them aspects of the social world outside of their normal experience and allowing them better to control it. Conceptualized thus, research helps to bring about social change only "from above," or as Silverman (1985) puts it, sociologists become "servants of the state."

In an attempt to modify the enlightenment model in order to meet some of these criticisms, Finch proposes first that researchers should aim to widen the policy debate by including groups outside the state or other powerful hierarchies in the research and dissemination process. Second, she suggests that researchers should be concerned with policy-making and change at *all* levels, including what she calls "the grass roots." This is seen as especially possible when qualitative research methods are used since these often involve close contact between researcher and researched. The notion of the detached investigator coming "from above" is already at a discount. Finch (1986, p. 231) writes:

> . . . one important implication of policy-oriented qualitative research is that it has the potential for engaging the researched as well as the researcher in evaluating the status quo and bringing about change . . . The model of using research, not to pass information upwards to remote 'policy-makers,' but to be used by the people researched . . . involves acknowledging that policy is not only implemented, but often 'made' and re-made at a very small-scale level.

I find myself in sympathy with the intention to research social issues in ways that democratize hierarchies of power rather than reproducing or maintaining them. This is not to say that the "grass roots" approach is either a panacea or its aims easily accomplished. On the contrary, it may be difficult to practice sociology along the lines suggested by Finch and a number of difficulties can be easily anticipated. At the most crude level social research is funded by the powerful who may not be entirely happy to see those over whom they exercise control involved in the process of social change. There is also the problem that social researchers may become, to put it in its most innocent light, the unwitting agents of the powerful by unconsciously adopting the interests and perspectives of those on whom they depend for funding. Nor would it be wise to assume that the grassroots are more receptive of ideas for social change than those with more power—conservative ideologies, for example, can traverse the hierarchies of an institution. Finally, it seems equally unlikely that, in parallel with other levels of policy-making, social research will have a direct effect at grass roots level and sociologists may have to be satisfied with an indirect and diffuse influence.

Despite these difficulties, the project of relating sociology to all levels of policy and practice seems to me an important one to attempt. From the perspective of a teacher of health workers it is essential. My students quite rightly want to feel that studying the sociology of health and illness will in some way, however distantly and indirectly, help them in their work. They can, again justifiably in my view, feel angry with sociologists who make no attempt to respond to this need. The problem, I suggest, is more than the private trouble of sociologists working in settings such as my own but is rather a public issue for sociology as an institution.

OVERCOMING THE SPLIT BETWEEN PURE AND APPLIED

In this section I want to discuss three related suggestions that address the problems developed in the paper so far, and which I think might make some contribution to their solution. They are: the use of a social drama approach; thoughts on rethinking the process of research dissemination; and the need to change the culture and organization of sociology.

The Analysis of Social Drama

In discussing his fieldwork for an investigation of health and illness in a small Italian town, Frankenberg (1984) suggests that a sociology that is social at all levels of its practice must attend to three tasks: first, illuminate the relationship between subjective experience and social process; second, locate its insights in the struggle for and contradictions of social change; and third, work alongside others at all levels who are concerned to bring about progressive social change. This stance had particular relevance of Frankenberg's study since it was carried out at a time (the early 1980s) when the Italian health services were undergoing an important and wide-ranging series of reforms with which he, his informants and his research collaborators were all concerned. Nevertheless, the approach he advocates has a much more general importance in its attempt to engage with social change processes but avoid the elitism of social change imposed only from above.

As part of this, and at the level of methodology, he suggests that sociologists, as well as surveying and questionnairing, pay attention to the analysis of what he calls the "dramatic incident." It is an approach taken broadly from the Manchester school of ethnographics, of which Frankenberg is a member, and more narrowly from the work of Turner (1968). In considering his work on Italian health care Frankenberg compares the surveys and questionnaires of social epidemiology with the sociological analysis of dramatic incidents. While the former aggregate the social from the experience of individuals who are left passive and objectified, the latter attempts to grasp the relationship between subjective experience and social process by sharing and entering into the flow of social life. Dramatic incidents are those intensified moments that interrupt that flow. In doing so they both allow experimentation with new forms of social relationship and raise consciousness about existing social relations. Examples might include a work accident, the opening of a new hospital, a visit to a doctor, a cliff hanging episode of a soap opera, and the birth of a child.

The classic use of the social drama framework is found in Turner's (1968) analysis of cults of affliction among Ndembu peoples. Here the sickness of an individual can be the occasion for ritual process that allows for the expression of grievances that find their origin in the structural tensions of Ndembu social relations, especially those around succession crises in village leadership. The cure of the sick individual involves the expression and resolution of social tensions either by reaffirming existing relationships or by instituting social change. Although Turner's analysis is specific to Ndembu society as he described it at the time of his fieldwork, the shape of the underlying process of social drama has, as Frankenberg (1987) points out, a wider analytical usefulness. He has suggested that this process has four phases:

a. a build up of a sense of breach
b. a crisis which results in a period of "liminold anti-structure" when social relationships can be re-thought
c. elements of the structure are re-ordered
d. restoration of the old structure or the establishment of a new one.

I find in my teaching of health workers that the concept of social drama has two applications. First, it helps them to gain more from their study of qualitative sociological research in the field of health and illness. Work such as that by Graham (1987) on women and smoking or Stimson and Webb (1978) on going to see the doctor are already powerful teaching materials precisely because they give such important insights into the point of view of their clients and patients. Placing this material in a social drama framework moves the discussion from the patients point of view and toward the patient's total field of social relationships.

This point relates directly to the second application that I have found in my own teaching. It is that with the social drama framework I can help health workers to analyze social aspects of their own practice. I have found that it has been directly helpful to, for example, the community nurses whom I now teach, helping them to structure their insights into the social dynamic of their patients' experience of sickness and its impact on their wider social network.

Rethinking Dissemination

For sociologists who have not worked on externally funded research the term "dissemination" can be a puzzling one. I remember, for example, speaking during a conference tea-break with a colleague well known for his theoretical writing. Our conversation turned to writing up our work on parenthood education and I used the term dissemination to describe the plans we had made. "Dissemination?," he asked, "You mean publishing?"

Did his unfamiliarity with the concept of dissemination signify the deeper split between applied and pure research? My own experience indicates that it does. In writing up my doctoral work, for example, I made little attempt to share my insights with people outside professional sociology, not only the particular parents, children and teachers of the school I studied, but also, at a more general level, pressure groups or other organizations concerned with the politics of education. My plans to publish from the research are confined entirely to journals of academic sociology and I had not thought of writing for a wider audience—or at least not until the invitation to contribute this paper prompted me to rethink the issue. I now see that there is a wider audience for whom my research could be useful and to whom I would like to address myself. Certainly there are topics dealt with in that work that are of wide interest: the pressures children experience in the transition to secondary school, despite the supposed end of the 11-plus; the contradictory pressures that schools impose on parents around decisions about sickness absence; the hidden curriculum of gender in teachers' handling of fitness, sickness and health.

Even so, having recognized the wider points of relevance of my research, there is something unsatisfactory about conceptualizing this as a matter of reaching "a wider audience." This implies a notion of dissemination in which the sociologist appears as the only sapient and active principle. The "audience" by implication are unknowing and passive, the receivers of the sociological insight rather than the source of it. This is profoundly elitist. It assumes that insight into social processes is in some way the prerogative of the professional sociologist. If such a conception is held, dissemination inevitably becomes, in a phrase borrowed from Habermas (1979), a distorted communication context in which critical knowledge cannot flourish. It may be that Habermas's ideal of open exchange and democratic dialogue is utopian but nevertheless, and at the risk of voluntarism, I believe that in our practice as sociologists we could go a good deal further in attempting to realize it than is currently the case. We could at least acknowledge that the conventional methods of dissemination, such as the lecture and the research report, encode a hierarchical relationship and begin to think of ways in which the barriers to open communication might be broken down.

From this point of view I now find that I am much happier with the social practice of sociology that was involved in our applied research. It brought us into contact with teachers and curriculum developers with whom we worked directly and because of this we were able to make some contribution to their construction of an anti-sexist and anti-racist curriculum.[7] Such contact changed the character of our research, especially in the dissemination phase. Instead of being a one-way transmission of our research to the practitioners it began to become a real exchange. We *learned* a great deal from teachers and curriculum developers during our attempts to disseminate. This included both new insights into the content of our work and their suggestions about how we could use more imaginative methods in the dissemination process itself. Central to this was our gradual adoption of a "workshop" format for dissemination activities. We were introduced to group work methods and approaches that eventually displaced the more traditional methods of dissemination such as lecture or talk on which we had initially depended. By the end of the project we had been in workshops with teachers, health visitors, the Family Planning Association and the National Childbirth Trust. During these we learned some of the ways in which research can be actively appropriated by other practitioners and used to focus their experience and ideas. We learned, and were taught, how to treat others not as an audience for our research but as co-participants in its interpretation.

Much of this (to us) new style of dissemination rested on the use of group work. Elsewhere in this volume Burgess, Goldsmith, and Harrison report how group work methods can be used as a highly effective method of data collection. We found that they are also useful in dissemination. By acknowledging the insights and knowledge that teachers and health workers brought to these sessions, by implicitly valuing them and facilitating their expression, "dissemination" of research became more of an exchange in which understandings and perspectives, including our own, could be changed. This allowed practitioners collectively to work out their own agenda of issues and questions through which they could appropriate whatever they found useful or insightful in our research.

It is significant too that one of the most stimulating of such workshop activities arose out of the analysis we made of the use of birth films in the classroom (Prendergast and Prout 1985). Such occasions are precisely incidents of social drama which fuse together the subjective and the social. The social relations on the screen—the unexpressed pain of a prone and naked woman, the male, controlling voice of medical commentary, the positioning of the camera such that it gives a viewpoint that no woman in labor or her partner ever sees, and the impersonal discourse of fetal development—all this intersecting with the social relations, including the gender relations, of the classroom made for a moment of dramatic social and personal precipitation.

We were able to use this material as the basis for a workship activity in which teachers and others could consider their approach to teaching about birth to secondary school pupils. Instead of hearing us tell them about our research we asked them to take part in an activity to which they could bring their already considerable understanding and experience but to which our research would be relevant. It consisted of them working in groups to write the outline of a film about birth for use in schools, a process that encouraged them to raise and explore issues such as the purpose, content and potential audience for such a film. Participants brought a range of perspectives to the task and were able to share these with each other. They were able to compare their ideas with those of others and also with a film commonly used in schools. The discussion that followed was invariably lively, productive, sensitive, and wide-ranging. Our research was only one contribution to the discussion but the activity created a context which made it more meaningful to the participants.

The openness of communication for which we strove in these sessions seemed helpful to the teachers and other practitioners who took part. It was certainly useful to us in the continuing analysis and interpretation of our fieldwork materials. This point underlines the more general methodological significance of our experience. The recognition that data collection and analysis are not two separate activities has become a commonplace of fieldwork methodology. The same suggestion has not, however, been fully developed in relation to the links between data collection, analysis and dissemination. The importance of checking interpretations with informants is often mentioned, but the issue now seems to me to be one of much broader scope than this alone. We need to consider not only returning to particular informants for their comments on our analysis and interpretation but structuring the wider *dissemination* of our work in such a way that it yields fresh data and insight that can be fed back into the research process.

There is also a parallel ethical point here. Just as some methods of data collection tend to place informants in a passive and depowered position (see, for example, Oakley's 1981 discussion of interviewing), so too with some methods of dissemination. The hidden curriculum of many "research presentations" to practitioners is that they are themselves incapable of insights into their own position and practice. To my mind this explains why so many such occasions are tense and unproductive. Group work methods do at least offer some ways of implicitly recognizing that practitioners too have valid insights that need to be placed on the agenda whenever research work is being discussed. In my experience this can help research to be engaged with and felt as empowering.

Changing the Culture of Sociology

Discussion of research, policy and practice tends to concentrate on the fate of research material once it has been fed into the decision-making processes of "user" organizations. In this respect the engineering and illuminative models are similar since neither pays much attention to the institutional context within which sociological research is itself organized and conducted. Earlier, I suggested that, from a researchers point of view, the social organization of sociology has an important bearing on the experience of research. In particular, in the lived culture of sociology a division is maintained between "pure" and "applied" work, a division which, at a personal level, I experienced as a split between the two types of research project on which I worked simultaneously. Like all such systems of opposition the terms pure and applied imply a deference as well as a difference (Derrida 1967). It was, perhaps, this implied hierarchy that led me to think of one area of my sociological work as "proper" than the other.

The division between pure and applied work can clearly be a useful one and I would certainly strongly resist the notion that all sociological work should be tested against some standard of "usefulness." It would in the first place be virtually impossible to reach a consensus on the meaning of utility. In the present political circumstances simply to privilege usefulness would invite philistinism at best and censorship at worst. It would also merely invert the existing hierarchy of theoretical and practical work—a move that I suggest would be ultimately self-defeating. Promoting practicality over theoretical purchase would spell the death of research the real utility of which lies precisely in its adequacy to comprehend the complexity of real social situations, just as surely as the one-sided dominance of theory leads to sterility in practice. The problem, it seems to me, is not so much to blur the distinction, or simply to invert the existing pattern of dominance, but to break down the pure/applied hierarchy within sociological culture.

I do not want to underestimate the distance that sociology has already moved in this direction and it may be that my own experience leads me to overdraw the picture. I fully recognize that many in the profession share a concern for work that is socially salient as well sociologically valuable. This high-tide of theory, or rather what Clyde Mitchell (1983) referred to as the mistaken view that theoretical work means only work on and with theories, has passed. Nevertheless, theory remains a dominant symbolic representation of purity in sociological work. Mary Douglas (1970, p. 191) has pointed out that living situations when symbols of purity are dominant:

> . . . (become) irremedably subject to paradox. The quest for purity is pursued by rejection. It follows that when purity is not a symbol but something lived, it must be poor and barren. It is part of our condition that the purity for which we strive and sacrifice so much turns out to be hard and dead as stone when we get it.

For some by choice, and others by force of circumstance, and still, alas, for some not at all, sociology is engaging with the profane world. Empirical work and applied work are back on the sociological agenda. But while we cling to old symbols of purity we run the risk of being like those social formations described by Douglas (1970, pp. 166-187) as wanting their cake and eating it: Enga men want to fight their enemy clans but yet to marry with their clanswomen; Bemba women want to be free and independent and behave in ways that threaten to wreck their marriages, and yet they want their husbands to stay with them. If sociology divorced itself from social policy and lived at loggerheads with empirical work, then a reconciliation contains the danger of recapitualting these old divisions—this time not through separation but by becoming a system at war with itself.

Assuming that such an outcome is considered to be undesirable, what can be done to prevent it? One line of approach might be to give greater acknowledgment to the applied research done in sociology departments and a good starting point would be to rethink the position of contract researchers. The marginalization that many experience, and which is reflected in my account, is a problem for the individual but also for the professions as a whole. It is also a problem that has become increasingly sharp as contract research has ceased to be mainly a bridge into teaching posts and has become a way of life for a whole swathe of practicing social scientists. It is further sharpened by the tendency for contract researchers to be as well qualified and have publications records on a par with many of their colleagues in tenured departmental positions. The solutions to these problems are partly trades-union ones concerned with improving the conditions of employment of contract researchers, but there are areas where what is crucial is the social organization and culture of the profession itself. Contract researchers, for example, need to be fully involved and welcomed into the government of departments, a move that will help to overcome that sense of being unwelcome guests at someone else's party.

A second line of approach is more difficult and deals with the more deep rooted problem of the relationship of theory and practice in sociology. It is, however, encouraging that recent writing (for example, Finch 1986; Silverman 1985) has begun to take up the challenge of integrating practical implications into sociological work. Finch provides a helpful model of a new relationship between qualitative research and policy that includes *both* involvement with practical issues, as they are defined at all levels of institutions, and an insistence on the central importance of theory for empirical sociology. I have already indicated my preference for the Manchester school approach to ethnography, especially the small-scale processual study and the analysis of dramatic incidents, as a means by which theory and practice can be brought together in sociological analysis. The example of birth films in the classroom illustrated one way in which this was attempted. Clearly, however, there are many other possibilities that deserve experiment and development.

Silverman has attempted to shift the terms of the debate by highlighting the shortcomings of one of its usual points of reference. He suggests that discussion on the policy implications of sociology still tends to take place around Becker's classic question "whose side are we on?" Each of the traditional answers (variously detached scholar, political partisan and servant of the state) have particular shortcomings and all, he suggests, are too ambitious. He suggests that a more fruitful and appropriately modest starting point is the question "What do we have to contribute?" I do not believe that choosing sides (or, to put it another way, politics) is so easily avoided. Neither should we allow the possibilities for choice to be restricted to those outlined by Becker, in particular, it may be possible to be political without taking on the rigidity and closed-mindedness implied by partisanship.

Nevertheless, the call to modesty seems to me to be timely in at least one important sense. It is that if we wish to focus more sociological attention onto issues of policy and practice then such work will bring us into contact with practitioners who daily struggle with the difficulties of their situations. As sociologists we frequently put ourselves in the position of criticizing such practitioners—almost to the point of it becoming a notorious stock in trade. Less frequently do we ask what we can learn from them. I am impressed by the efforts of many of the teachers and health workers[8] with whom I have had contact through my research in their struggle to unite political consciousness, personal experience and professional theory. In particular, many seem capable of a level of critical self-reflection on their individual and collective practices which, notwithstanding excellent collections such as those edited separately and jointly by Bell and Roberts (1981, 1984), put sociologists to shame. I end, therefore, on the unfair, incomplete, inadequate but I think not unhelpful suggestion that by comparing their experience with ours we might learn from them how to be better practitioners ourselves.

NOTES

1. This paper was originally given at the ESRC Fieldwork Methods Seminar on Qualitative Methods and Policy, University of Warwick, March 20, 1987. It was repeated at a similar meeting at the University of Edinburgh, June 12, 1987. I am grateful to the seminar participants for the discussion that followed on each occasion and to Bob Burgess for his helpful comments.

2. In March 1987 the Health Education Council, a quasi-independent body funded by the Department of Health and Social Security, was abolished and replaced with the Health Education Authority, also funded by the DHSS but constituted as a special District Health Authority. This change was widely interpreted as one designed to bring health education under greater direct governmental control.

3. None of this is to imply that antenatal care is either effective or efficient in reducing infant mortality and morbidity, or affecting maternal health. Indeed this has been widely questioned (e.g., Kerr 1980; Hall 1981). Our research suggested that schoolgirls were already convinced of the benefits of antenatal care before any specific educational efforts in school. In some ways they had a too optimistic expectation of what antenatal care can achieve and further simple propaganda

messages would seem to have little impact. We argued for educational approaches that treat the topic of atenatal care with some recognition of its complexity (see Prout 1985a).

4. The situation has shifted somewhat since I first made this point, for example, Silverman (1983). It was also not true of the earlier literature on child hospitalization (e.g., Hall and Stacey 1979). A discussion of children and childhood in medical sociology is found in Prout (1987).

5. The advent of AIDS seemed to radically change the character of the political debate on sex education.

6. A right wing moral-political lobby now known as Youth and Family Concern.

7. For example, Braun and Eisenstadt (1985).

8. Examples I have in mind include: Lesley Smith's work on the negotiated curriculum (Smith 1986); Hazel Slavin's work on training for AIDS Helpline volunteers (Slavin 1987); and the structure and process of the Diploma in District Nursing course at the South Bank Polytechnic, designed by my colleague Kate White, and the related research she is doing on the transition of her students into practice.

REFERENCES

Bell, C. and Roberts, H.
1984 *Social Researching: Politics, Problems and Practice.* London: Routledge and Kegan Paul.

Bluebond-Langner, M.
1978 *The Private Worlds of Dying Children.* Princeton, NJ: Princeton University Press.

Braun, D. and Eisenstadt, N.
1985 *Family Lifestyles and Childhood.* Milton Keynes: Open University Press.

Bulmer, M.
1982 *The Uses of Social Research.* London: Allen and Unwin.

Campbell, G.
1984 *Health Education and Youth.* Lewes: Falmer Press.

Cornwell, J.
1984 *Hard Earned Lives.* London: Routledge and Kegan Paul.

Davies, J., Prendergast, S. and Prout, A.
1978 *The Health Knowledge of Children. Their Parents and Teachers.* London: Health Education Council (Mimeo).

Derrida, J.
1967 *L'Ecriture et la difference.* Paris: Seuil.

Douglas, M.
1970 *Purity and Danger: An Analysis of Concepts of Pollution and Taboo.* London: Pelican Books.

Finch, J.
1986 *Research and Policy. The Uses of Qualitative Methods in Social and Educational Research.* Lewes: Falmer Press.

Frankenberg, R.
1984 "Incidence or Incidents: Political and Methodological Underpinnings of a Health Research Process in a Small Italian Town." In C. Bell and H. Roberts, *Social Researching: Politics, Problems and Practice.* London: Routledge and Kegan Paul.

Frankenberg, R.
1987 *Aids and Social Anthropology.* University of Keele, Centre for Medical Social Anthropology (mimeo).

Graham, H.
1987 "Women's Smoking and Family Health." *Social Science and Medicine* 25(1): 43-56.

Gorbutt, D., Bowden, T., and Pring, R.A.
1972 "Education as the Control of Knowledge." *Education for Teaching 89.*
Habermas, J.
1979 *Communication and the Evolution of Society.* Boston: Beacon.
Hall, D. and Stacey, M. (Eds.)
1979 *Beyond Separation: Further Studies of Children in Hospital.* London: Routledge and Kegan Paul.
Hall, M.H.
1981 "Is Antenatal Care Really Necessary?" *Practitioner* 225:1263-65.
James, N.
1984. "A Postscript to Nursing." In C. Bell and H. Roberts, *Social Researching: Politics, Problems and Practice.* London: Routledge and Kegan Paul.
Kerr, M.
1980 "The Influence of Information on Clinical Practice." In I. Chalmers, *Perinatal Audit and Surveillance.* London: Royal College of Obstetricians and Gynaecologists.
Mitchell, J.C.
1983 "Case and Situation Analysis." *Sociological Review* 31:187-211.
Mills, C.W.
1969 *The Sociological Imagination.* New York: Oxford University Press.
Oakley, A.
1981 "Interviewing Women: A Contradiction in Terms." In H. Roberts (ed.), *Doing Feminist Research.* London: Routledge and Kegan Paul.
Prendergast, S. and Prout, A.
1980 "'What will I do . . .?' Teenage Girls and the Construction of Motherhood." *Sociological Review* 28(3):517-535.
1985 "The Natural and the Personal: Reflections on Birth Films in Schools." *British Journal of Sociology of Education* 6(2):173-183.
1987 *Knowing and Learning About Parenthood.* London: Health Education Authority.
Prout, A.
1984 "Teenagers and Motherhood: Who's being Naive?" In G. Campbell (ed.), *Health Education and Youth.* Lewes: Falmer Press.
1985 "Science, Health and Everyday Knowledge: A Case Study in the Common Cold." *European Jorunal of Science and Education* 7(4):399-406.
1985b "Teenage Girls' Knowledge of Antenatal Care and Its Implications for School-based Preventive Strategies." *Health Education Journal* 44(4):193-197.
1987 "An Analytical Ethnography of Sickness Absence in an English Primary School." Unpublished Ph.D. thesis, University of Keele.
1987 "Off School Sick: Mothers' Accounts of Negotiating Sickness Absence with their Children." Paper to the BSA Medical Sociology Group Conference, University of New York, September 1987.
Roberts, H. (ed.)
1981 *Doing Feminist Research.* London: Routledge and Kegan Paul.
Scott, S.
1984 "The Personable and the Powerful: Gender and Status in Sociological Research. In C. Bell and H. Roberts, *Social Researching: Politics, Problems and Practice.* London: Routledge and Kegan Paul.
Silverman, D.
1985 *Qualitative Methodology and Sociology.* Aldershot: Gower.
Slavin, H.
1987 "Training for the National AIDS Helpline." *Health Education Journal* 46(2):62-64.

Smith, L.
1986 *Dimensions of Childhood: A Curriculum for 16 Plus*. London: Institute of Education.

Stimson, G. and Webb, B.
1978 *Going to See the Doctor*. London: Routledge and Kegan Paul.

Turner, V.
1968 *The Drums of Affliction*. Oxford: Clarendon Press.
1977 "Variations on a Theme of Liminality." In S.F. Moore and B.G. Myerhoff (eds.), *Secular Ritual*. Assen: Van Gorcum.

Turner, V. and Turner, E.
1978 *Image and Pilgrimage in Christian Culture*. Oxford: Blackwell.

Young, M.F.D. (ed.)
1971 *Knowledge and Control*. London: Collier-Macmillan.

CONVENTIONAL COVERT ETHNOGRAPHIC RESEARCH BY A WORKER:

CONSIDERATIONS FROM STUDIES CONDUCTED AS A SUBSTITUTE TEACHER, HOLLYWOOD ACTOR, AND RELIGIOUS SCHOOL SUPERVISOR

Norman L. Friedman

This chapter considers a number of general, ethical, and procedural methodological points about what will be called "conventional covert ethnographic research by a worker." It does so in large part in relation to reflections on three such studies conducted by me in recent years: as a high school substitute, a Hollywood actor, and a religious school supervisor.

"Covert" research refers to investigation in which the research observer does not reveal to the subjects his/her true identity and/or that he/she is conducting a study. "Conventional" research done by a worker means that the research setting involves a relatively "conventional" or ordinary occupation, job, and/or

Studies in Qualitative Methodology, Volume 2, pages 189-204.

ISBN: 1-55938-023-3

organization, that is, one that is neither controversial, deviant, injurious, extreme, nor extraordinary. (Of course, it is possible for an essentially conventional type of work to demand unconventional behavior from a worker/researcher, but the ethical covert researcher can then choose to leave or resign if necessary.)

Covert research, of course, has the great strength of enabling the researcher maximally to participate in and observe natural social life without the unnatural reactions or alterations that subjects might display or fashion if they knew he or she was a researcher. When covert research takes place in conventional work situations, though, the eventual ethnographic report, given customary safeguards—such as the use of pseudonyms, changing minor details that might expose or be injurious, delay of publication until many workers have departed—usually will not be harmful or distasteful or threatening to the subjects. In truly conventional work settings, the fears and concerns of subjects are so low or even non-existent that the answer to the following question the researcher might contemplate would usually be *no*: If those you work with happened to discover that you were also a researcher secretly conducting a study of their job or setting, would they be shocked, angry, indignant, and demand that you stop?

THE COVERT RESEARCH LITERATURE

Much of the literature about covert research in general has viewed it as ethically and procedurally abhorrent. Critiques have dwelled on dramatic or extreme cases of (non-employee) covert research on the deviant or exotic—such as sectarian groups (Festinger, Riecken, and Schacter 1956), pseudo-patienthood (Caudill, Redlich, Gilmore, and Brody 1952; Rosenhan 1973), homosexuality (Humphreys 1970), alcoholism (Lofland and Lejeune 1960)—or on more controversial heavily power-wielding or coercive occupations, such as the police (Holdaway 1982) and the military (Sullivan, Queen, and Patrick 1958). Criticisms have included contentions that such covert researches are potentially or actually harmful and injurious to subjects, damaging to the research reputation of sociology, and/or questionable as to scientific accuracy (Erikson 1967). Covert researchers, it has been argued, lack informed consent of subjects, invade their privacy, are deceptive, and are sometimes methodologically unnecessary. They unethically "misrepresent" the researcher's intentions and/or identity (Bulmer 1982).

Conventional covert ethnographic research by a worker, though, seems to have escaped some of these stronger condemnations, especially when the sociologist/worker is seen as legitimately eligible and qualified for the work (rather than seen as an ineligible or unqualified imposter, fraud, or planted "spy"). For example, Erikson (1967, pp. 372-373), though critical in general of covert research, commented that:

> some of the richest material in the social sciences has been gathered by sociologists who were true participants in the group under study but who did not announce to other members that they were employing this opportunity to collect research data. Sociologists live careers in which they ... occasionally take jobs as steel workers or taxi drivers.... It would be absurd, then, to insist as a point of ethics that sociologists should always introduce themselves as investigators everywhere they go and inform every person who figures in their thinking exactly what their research is all about.

Among the many such studies done over the years, three well-known classics are Becker's (1951, 1953) on dance musicians, Davis' (1959) on cab drivers, and Dalton's (1959) on managers.

Though I will still call them covert, Whyte (1984, pp. 30-31) has even suggested that such studies are neither covert nor overt but "semiovert."

> In the semiovert role, you combine a regular job in the organization with your study of that organization. Since you appear to be earning your living like everyone else, full explanations of your research are not necessary.

In any event, the assumption is made in this paper that conventional covert ethnographic research by a worker has a traditional and acceptable place in sociology, and, though more rare than overt non-employee studies by outsiders, has been and should continue to be a legitimate and valuable research alternative for participant observation.

THE THREE STUDIES

Such conventional covert ethnographic research by a worker was the case in three recent studies I conducted as a high school substitute (Friedman 1983), a Hollywood actor (Friedman 1990), and a religious school supervisor (Friedman 1987). In all three, as worker/researcher, I intended to try to do field research while also performing paid job tasks. I intended in the future to publish research results when and if possible. In all three I did not inform those in the settings that I was conducting research while also serving as a fellow job seeker or employee, so as not to alter their "natural" behavior.

In the cases of the substitute and actor studies, being able to do the tasks satisfactorily, as "just another worker," was a part of the research on occupational role adaptations. This was similar to Hilbert's (1980, pp. 69, 71) covert research on being a student in an elementary teachers credential program. Hilbert did so in order to determine what it took to manage and sustain competent membership in that conventional group, in regard to "the mundane realities of day-to-day social existence" (p. 53). As he put it:

> Managing the secret *is* the research, *is* the display of typical membership. ... The analyst *is* a typical member by virtue of successful passing.

My research on a Jewish Sunday school, conducted while I was employed in the administrative and coordinative job of "Supervisor," in contrast, was more of a holistic organizational analysis of one setting, a "school ethnography."

High School Substitute

The research as a public high school substitute teacher was conducted during the 1979-1980 academic year in the high schools of three diverse school districts in Southern California. During an academic leave from my position as a university sociology professor, I worked as an officially qualified and certified substitute teacher in various schools for 45 days over a 10-month period. Though on district applications I indicated I was a Ph.D. (a bachelor's was the minimal degree required), I did not reveal on the applications that I had a regular ongoing university level professorial position from which I was on leave, nor in the schools did I use the "doctor" title, because I wanted, fo research purposes, to be viewed by adults and youngsters alike as "just anothe sub." Actually no person during the year asked me what other job, if any, had. (They were preoccupied with interaction with me as "just another sub," probably assuming that substituting was my sole job anyway.) And I did no indicate to anyone that I was, in addition to working as a substitute teacher also conducting sociological research about that occupational role. After brie pre-service socialization by the districts, I was called in to do the work in th schools.

Hollywood Actor

The research as a Hollywood actor, seeking and sometimes employed in job in films, television shows, and television commercials, was conducted fron 1977-1986. I originally wanted to try to study the occupational and cultura sociology of Hollywood not through overt interviews by an outside researche but by somehow injecting myself into the industry and art as a natura participant. I did this by declaring myself to be "an actor" (an occupation which strictly speaking, has no absolute minimal formal educational or experienc qualifications, at least about which everyone can agree, and really only require that the actor effectively do the job, if hired), and then searching for jobs an sometimes actually getting and doing them. Since actors were notorious unemployed and underemployed, this infrequent work practically neve interferred or conflicted with my schedule as a university professor.

I fared reasonably well in establishing credibility as a new actor starting ou and trying to succeed. This was partly because I had had some acting interest and experience earlier in life during the elementary and high school years in my Midwestern home town. And even though I had no similar experience for about the next 23 years, from ages 18-41, once into this study I applied myself step-by-step just as any other novice actor would.

Thus, along the way, I became involved in activities necessary to try to secure employment: had professional photographs made and resumes prepared, sought and secured agents, was admitted to unions, attended workshops, and sent mailings to casting directors. In the earliest jobs I was able to get, I worked as a non-union "extra" in a number of films, had a speaking role in one non-union film and a "principal" (major) role in one non-union commercial. Subsequently as a union actor, I eventually had numerous jobs as a "principal" speaking actor in television shows and a film, as well as major parts in commercials. Once hired, I did the job as skillfully as possible.

So as to allow others to speak and act naturally when with or near me, I did not indicate to anyone that I was, in addition to seeking and occasionally performing acting jobs, also conducting sociological research about that occupation and its milieu. At first I was concerned that my identity as a sociologist/professor, if revealed, might adversely affect or impede the establishment and maintenance of my identity as an actor. Eventually I decided that I would try to act like "just another actor" in Hollywood places, but would answer honestly if and when someone directly asked what "else" I worked at, though I would not otherwise voluntarily mention it.

I found (to my surprise) that when I occasionally did identify my "other job" (usually as "college teaching," and, if asked further, "of sociology," and if pressed still further, "at Cal State L.A.") to another actor or agent or a casting director, it made no difference in any way. It did not seem to alter their demeanor toward me, as it often would with people in other walks of life (usually making them decidedly more or less respectful). It did not seem to affect our particular means/end relationship. This was perhaps because so many people in the Hollywood entertainment industry are preoccupied with their own little social world, as all-important and all-consuming to them. Also, it is not unusual for most actors to have some "other job" or jobs, for income purposes. Moreover, a number of actors "moonlight" around Hollywood as acting *teachers*, so at least my being a teacher or professor did not sound so strange—even of sociology.

Religious School Supervisor

The conventional covert research at a private Jewish Sunday school was conducted by me when I was the supervisor there, 1980-1984. During that period I was employed on Sundays as the Supervisor of the primary grades (kindergarten-to-third) department of the larger religious school of a Reform Jewish temple in the Los Angeles area.

The job of Supervisor mainly required that I carry out the goals, policies, and practices of the School Director, and that I more immediately deal with whatever operational and instructional problems came up every Sunday. Unknown to others, I was also covertly doing ethnographic research about the organization, and the Supervisor job turned out to be an excellent vantage point from which to observe a wide array of activities and interactions every Sunday. It routinely and legitimately took me in-and-out of classrooms, the front offices, private conferences and meetings with the School Director, faculty, students, and parents, and special activities and gatherings. (It also gave me access to pertinent documents, such as bulletins and memos, curriculum outlines, learning materials, and so forth.)

The board and administrators who hired me were aware of my full-time "other job," as a university professor of sociology (and a former department chair). Indeed that identity and experience, along with some background in Judaica, were what in large part led them to feel that I could handle the job of Sunday Supervisor. Once on the job, the teachers under my supervision were informed that I was also a university professor.

Thus, as discussed above, in all three case studies the fact that I was conducting covert but conventional ethnographic research was not revealed to fellow jobholders or seekers. The fact that I was also a university sociology professor was differentially dealt with to fit each somewhat unique research circumstance and its imperatives: from concealment and non-indication as a substitute (so as not to be seen as somehow overqualified or extraordinary in the role), to provision of information about it as a Hollywood actor, but only when directly asked to (out of the fear that an overemphasis on it would divert attention from me as an actor), to full disclosure and emphasis on it as a religious school supervisor (as a desirable qualification for the job and a source of recognition from and authority among co-workers).

These somewhat different approaches were taken to maximize research-related operations and outcomes, as well as sometimes to facilitate job acquisition and/or performance. In these regards, perhaps the following comment by Roth (1962, p. 284) is appropriate:

> social science research cannot be divided into the "secret" and "non-secret." The question is rather how much secrecy shall there be with which people in which circumstances. Or, to state the question in a more positive (in more researchable) manner: When we are carrying out a piece of social science research involving the behavior of other people, what do we tell whom under what circumstances?

Therefore, a general point here is that: Which facts about the research or the researcher's identity are treated as covert or secret, and in what ways, will need to differ according to research goals and needs, job-related imperatives and circumstances, or both.

OPPORTUNITIES AND ADVANTAGES

This type of research affords some special opportunities and advantages to the researcher. Quite obviously and candidly, it enables the researcher to conduct important research while also earning income. In a sense, the income can serve as a sort of paid "research grant," but without some of the special explicit or implicit obligations to or "strings attached" by research grant agencies, such as the government, foundations, corporations, or the home university. For instance, it is usually easier, if necessary, to quit the job than it is to quit the government, foundation, corporation or university grant.

Graduate students in sociology and related fields have, of course, been known to combine work and research, overtly, or covertly. For instance, a graduate student in anthropology who worked, partly covertly, as a cocktail waitress in order to study that role, wrote that:

> A third issue complicated the situation in my case: my role as an employee and the fact that the research project was initially secondary to this role. The problem is becoming more frequent as anthropologists turn to their own society for study and research funds become increasingly scarce, compelling graduate students to take jobs, both as a means of support and as a place for research (Mann 1976, p. 99).

To be sure, such income-related needs are not limited to graduate students, but can often extend upward to their non-affluent professors as well.

Advantages of worker research, though, are not limited to the financial. There can be other personal gratifications and satisfactions, such as in learning new skills, doing something different from college teaching, and/or making a contribution to another field. In the cases of my studies, substitute teaching provided the challenge of eventually being asked to teach almost any subject field in which there was an opening on a given day, and then attempting to do so satisfactorily, and religious school supervision provided the opportunity to make an "in the trenches" personal contribution to Jewish education, long a field of interest to me.

Hollywood acting provided the challenge of trying to "make it" in an exceptionally competitive field, working at various hierarchical levels (lowly "extra" to more pampered "principal"), and flexing creative muscles and inclinations. In contrast to "professing," a generally private, isolated, and sober/serious set of activities, Hollywood acting provided more interaction with (usually) less pompous and solemn types who were more able to laugh at themselves and not take themselves so seriously, unlike academicians. I frequently noticed at television shows and commercials, after some ridiculous line or action (say, gobbling up a pizza), that an actor or director would say something like, "Well, it sure ain't Shakespeare!" And I wondered how many academicians would say something similar, so refreshingly honest, after completion of one of their class lectures.

But the difference that most struck me was the collaborative group nature of being in a film, television show, or commercial, as opposed to the individual and private life of college teaching. This was revealing to experience. In most instances in college teaching you have sole responsibility for almost everything that you do in the classroom: planning, writing, and performing. In the mass media, I always had a sense of being a contributor, in whatever capacity and to whatever extent, to a collective effort and product. That feeling was a gratifying change-of-pace from the privatism of college teaching. It was especially strong when working on the television situation comedies that were taped in front of a studio audience (such as "Facts of Life" and "Newhart"). I had the definite sense of working as part of a team—in conjunction with other actors, the director, producer, writers, technical crew—to present the episode to the studio audience. And when the taping was over, I shared in the sense of group elation, disappointment, or whatever, about the results. As I left, I told some others goodbye and said, "Nice working with you—hope to work with you again," a common cultural saying in that setting.

A SPECIAL ETHICAL OBLIGATION

The conventional covert researcher/worker has a special obligation, beyond the customary ones to subjects, such as confidentiality and non-injury. He/she has a special obligation to perform the work in as highly skillful, effective, and successful a manner as is humanly possible. The fact that the researcher has been hired and is being paid for services rendered, as a worker, means that those responsibilities must take precedence over any other considerations, including very often the covert research activity itself. The employee must deliver the full measure of effort. Not to try to do so is what would definitely be unethical in this type of research. In his/her dual role, how well the worker performs the job role, rather than any secrecies maintained in the role of researcher, *is* the central *ethical* issue of importance to and for others in these conventional settings. Not to try to do so is what would definitely be unethical in this type of research.

Assuming a new job is often difficult and time consuming in regard to mastery of tasks. Initial (and sometimes ongoing) adaptation efforts often are quite challenging. Mann (1976, p. 100), for example, almost quit her research-related work as a cocktail waitress, in response to the initial difficulties of its demands and expectations, including sexism. I faced pressing problems of "order-maintenance" and "assignment-execution" that needed to be grappled with as a new substitute teacher. Student "discipline" problems that I needed somehow to deal with, to one degree or another, including talking when I had asked the student not to talk, throwing objects, not following directions about assignments during class, ignoring my presence while doing whatever else the student wanted to do, moving around the classroom without permission, answering the roll call for another student, shouting profanities, standing outside the classroom rather than entering, acting defiant or sullen, and verbally and/or physically fighting with another student.

One special difficulty was what I have called the "phenomenon of the false friend." In many of the classes I substituted in, a recurring phenomenon transpired. Slightly before or at the beginning of class, some student would come up, act very friendly, and ask if he or she could help. Often the student would say something like, "Watch out—this is an awfully bad class—it's terrible!" Almost invariablly, this seeming helper and/or advice-giver would turn out to be a "false friend," that is, he or she would turn out to be the main class discipline problem, rather than part of the solution to any problem. The warning turned out to be about the "friendly" student, who became in turn the major saboteur of classroom order rather than its ally.

As a Hollywood actor, I had to "learn the ropes" of trying to get ahead and concentrate on successful career strategies toward that end. Like some other new would-be actors, this involved some initial experience in work as a movie and television "extra" or "atmosphere" background non-speaking player until "better" acting jobs materialized. Such low-level extra work soon became frustrating, because of the lack of status and authority, and the boredom, monotony, and impersonality. It became demoralizing for those would-be actors, like me, who aspired to better and higher acting jobs. As one remarked: "You're close to what you want to do, but *you're* not doing it."

Initially as a Sunday school supervisor, I was faced with the problems of smoothing out and filling gaps in my knowledge about Jewish primary grades, students, curricula, learning materials, and approaches to innovative instruction. Less experienced teachers asked me about the nature and advantages and disadvantages of different types of approaches to "discipline" (or "classroom management" styles) for primary grade youngsters, and I needed quickly to read in more detail about such matters in order to respond and supervise satisfactorily. Also, because about 25% of the students were from Jewish/Christian intermarriages, I was pressed early by teachers concerning how they should react and respond to comments from their students that "I'm half and half" or "We have Hanukkah *and* Christmas" or "We're Jewish *and* Christian."

Thus in all three studies, initially mastering and effectively performing the job tasks were often difficult and time consuming. Nevertheless, doing so constitutes a special ethical obligation in conventional covert ethnographic research by a worker.

RELATED PROCEDURAL CONSIDERATIONS

Such realities also mean that conventional covert ethnographic research by workers, all other things being equal, usually will (and should) take longer to complete than overt ethnographic studies by outsiders. The overt outsider researcher is better able to give full focus and attention to observation, while the covert inside worker must give time to job participation, often in priority over and to the detriment of research observation. Faced with two tasks, the outcome will usually take longer, although the eventual results are often enriched by the double-duties.

These dilemmas are intensified by the practical difficulties of trying to work and conduct research at the same time. As Holdaway (1982, p. 171) remarked:

> establishing a covert research role involves a constant process of self-reflection. ... There were occasions when I forgot that I was researching. ... There were times when my administrative duties were spoiled because I was involved in the minutiae of police action. Covert research involves a constant heightening of sensitivity to the possibilities of recording conversation, action or whatever—such activity is exceedingly demanding and can, after a time, become stressful.

Similarly, being a Hollywood actor was frequently a painful experience that sometimes prevented my greater attention to research-related thoughts, and, therefore, necessitated a longer-than-intended study. This was a considerably difficult participant observer role. Once I thrust myself into the role of would-be actor, it became very emotionally engulfing and draining. I could not help but also dream the dreams and feel the failures. Since my career path heavily involved auditioning for television commercials, I had to live that life of frequent auditions, pre-audition anticipation, then wondering how I did, some "callbacks" for second and third auditions, then wondering if I got it, and feeling rejected when I almost always did not. I not only, of necessity, studied rejection, but actually lived it as well. While I tried to cope with it as well as the next actor (for instance, by "looking ahead, never backward"), cumulatively it was taxing. Also, it prompted strains in home life. My wife eventually could recognize when I had not gotten a job sought; I behaved in a more irritable and depressed fashion. Ups and downs. Highs and lows.

There are some practical research techniques that try to cope with these double-duty dilemmas. Hilbert (1980, pp. 65-68) recalled that he was able to carry around a notebook in which he visibly took notes as a covert researcher while appearing to take notes in the role of elementary teacher credential student. I similarly carried a weekly schedule as I walked around to the various classrooms and offices as the religious school supervisor, on the back of which I frequently jotted down important field notes (but not at the expense of job performance). I also usually had the appropriate writing articles around to make some at-the-time notes as a substitute, but obviously this was less the case as a Hollywood actor, and those notes needed to be made away from actual interaction settings, for the most part.

Another common technique in covert worker research is the natural job-related and casual "conversation" with co-workers and/or clients that also cautiously probes for research information (Dalton 1959, p. 280). The researcher must be careful not to probe or question so much or so obviously that it arouses suspicions about his or her research identity or intentions. The following was a light conversational probe:

Co-Worker: Three years ago I had four national commercials running at the same time!

Researcher/Worker: Oh-yeah—that's great. And, er-why haven't you had any since?

I used this conversational interview technique in all three studies, but sparingly, partly because of its exposure dangers, and partly because used in excess it diverts too much from the natural course and flow of interactional relationships. (It is almost impossible, though, consciously to avoid doing it altogether.)

COMPARING COVERT PARTICIPANT OBSERVATION WITH OVERT IN-DEPTH INTERVIEWING

This conventional covert participant observation provided me with some interesting contrasts with my earlier qualitative research experiences that used overt in-depth interviewing in studies of the careers and tasks of community college teachers (Friedman 1967, 1969) and the ethnic/religious orientations of Jewish professors in four-year colleges and universities (Friedman 1971, 1973). The overt in-depth interviews in those studies allowed me better differentiation of types and variations of interviewees and their orientations and feelings. But the three covert participant observation studies better enabled me empathically to understand, through my similar experiences, the meanings and depths of orienations and feelings, as well as more fully to comprehend how and why certain conduct develops and transpires *in situ,* especially the ways workers manage and grapple with their tasks and engage in actual behavior that differs from ideal or advocated behavior.

Thus the community college teacher interviewees were divided into important "pre-organizational career types"—High Schoolers, Profs, and Grad Students—and I showed how their organizational careers, special orientations, and task adaptations flowed from these variations. The Jewish college and university professor interviewees were typed as to differences in stances toward the larger Jewish community, as Insiders, Fellow Travellers, Indifferent Separatists, and Renegade Separatists.

While the three covert participant observation studies did not uncover quite as rich typologies and vivid images, they did afford a much greater feeling of more "certain knowledge" as to how participants became engulfed in situational demands and attempted to negotiate and manage them. In contrast, there is usually some skepticism about interviewee accounts of events, no matter how vivid and how recurrent they are. The accounts by community college teacher interviewees as to how they adapted to task demands, for example, was knowledge somehow not as compelling, real, or "certain" as actually having been a high school substitute teacher and having experienced *in situ* the difficulties of adapting to those tasks. More specifically, the regular high school teachers often provided very sketchy (and even non-existent) "lesson plans" for the substitute, which required the substitute to engage in quick, on-the-spot content mastery, improvisation, and the development of some class content time-fillers. In English classes, my stock fill-in items became popular song lyrics viewed as poetry and impromptu oral quizzings about how certain words are spelled. In addition, the emotional feelings expressed by overtly interviewed respondents never approached the depths of emotional empathy acquired in participation in the ups and downs of the successes and failures of the Hollywood actor.

Another area where covert participant observation excelled over overt in-depth interviewing was in being better able to identify the discrepancies between the ideal and the real, between what were professed or advocated in contrast with ideas and behavior that were actually held and performed. For substituting, I initially wondered how substitutes could step in at the last minute and effectively teach such a wide variety of specialized high school subjects. I eventually found that this was manageable because the regular teachers' classroom directions frequently called for students independently to be working on some writing or reading (or artistic or scientific) exercise, or to be taking an examination, and/or to be viewing a film or videotape during some, most, or all of a classroom period. Such kinds of independent student activities, therefore, filled up much of the allotted class time, and in turn freed substitutes from having to possess and display more wide-ranging knowledge.

For the actor, occupational culture norms urged setting a reasonable "time limit" on success searching and to have a career "plan of action." In actuality, as it turned out, most actors ignored such rational advice. And while actors frantically searched for jobs and success, they also developed ways of coping with rejection and failure, such as urging themselves and others to "count your blessings":

> I say stop wallowing in self-pity that you're not going to make it. Who said life's going to be easy? Nobody asked you to become an actor. Sad but true, nobody wants to hear your problems. If you get emotional and think life is over for you, take a trip to the Children's Orthopedic Hospital in downtown L.A. right off the Harbor Freeway, and go in there and say, "I'd like to be a volunteer for a couple of days a week." Take a walk through those wards and then come out of there and tell me how unfortunate your life has been.

In the Jewish Sunday School, covert participant observation uncovered ideal/real differences in curriculum and instruction. Thus, while the formal curriculum guide for grades kindergarten, first, second, and third included the study and celebration of Jewish holidays as just one subject, in reality, for various reasons, attention placed on Jewish holidays actually informally monopolized the class time available far beyond its place and time allowance in the formal curriculum. Also, though teachers were presumably hired because of their special qualifications, their various background deficiencies in Jewish education meant that the School Director in reality actually needed to give them extensive guidance, advice, materials, suggestions, and direction for their teaching. As she privately communicated: "We do almost everything but teach that one or so hours for them!" The above were, therefore, some of the differences and discrepancies in ideal versus real behavior, uncovered in the covert participant observation studies, that would have been less likely to have been discovered through overt in-depth interviewing.

GIVING SOMETHING BACK

Conventional covert ethnographic worker research can also make policy/practice contributions to the work-related areas, thus "giving something back," especially through publication and dissemination of research findings. In the cases of the substituting and supervising studies, I deliberately sought and secured publication in practitioner-oriented journals, *Urban Education* and *Jewish Education,* respectively. This was done in the hope that the findings might be suggestive and/or helpful for the practitioners and policymakers who read those publications. It was a way to try deliberately to "give something back" to the type of work or setting, in exchange for the advantages afforded me of covertness.

In the case of Hollywood actors, a colleague and I (Friedman and Friedman 1987, pp. 289-291) attempted to show in an article about the practice concerns of career and employment counseling, which appeared in *The American Sociologist,* that the job and success search activities of actors definitely have implications for the job and success searches of other workers:

> Because job search takes an "extreme" form in the occupation of screen acting—constant search and rejection—it perhaps has something more general to suggest ... about the more "ordinary" job/success search imperatives of other occupations. One of these suggestions ... is in the common types of sayings urging job search staying power in the face of adversities: "hang in there" or "remember, perserverence and persistence!" ... Actors have long been in the business of trying to get jobs where too few exist, and the drive, determination, care, cultural ideas, and behavioral adaptations they utilize in pursuing their goals might be suggestive and even worthy of emulation by others more recently experiencing employment difficulties.

In the case of the religious school, I assumed that Jewish education practioners and policymakers, reading the descriptive analysis in their major professional journal, *Jewish Education,* would draw some of their own policy/practice conclusions. But in the case of high school substitute teaching, I went a step further to spell out the implications of my work/research experiences, in the belief that substitutes, as teaching practitioners, have a special contribution to make to regular classroom education:

> there is, to be sure, a higher level of assignment—execution that sometimes occurs. Most substitutes, because of their unique individual sets of life histories and experiences, are capable sometimes of conveying to students knowledge, insights, and understandings that are better than, or at least different from, those of the regular teacher, thereby enriching the regular instruction. I recall some of those moments in my own substituting, when such enrichments seemed movingly to take place in clarifictions, comments, asides: talking to an English class about seeing *Death of a Salesman* several times over the years and how that affected my changing definitions of the meaning of the theme of "success"; while discussing the film *The Story of a Writer* ..., indicating how teenagers can be creative too, or help to develop creativity in younger children even during such activities as babysitting; commenting on points high school history books often omit about the history of American minority groups; explaining to ninth graders how and why Shakespeare expressed romantic sentiments of youngsters, in *Romeo and Juliet,* so much better than most of us can; explaining to a "Fantasy/Adventure" literature class why, psychologically, genre fiction is so popular with many people, something they had not considered before all during the semester. The point is not that I said things that were so special, but that *most* substitutes have unique personalized touches to add to regular instruction in the course of their assignment executions (Friedman 1983, pp. 124-125).

SUMMARY

"Conventional covert ethnographic research by a worker" has an established and valuable place in sociological research. It has the advantages of covert research without many of the disadvantages of covertly studying more unconventional settings.

In three such researches—as a high school substitute, Hollywood actor, and religious school supervisor—co-workers were not told that I was also conducting research. However, varied stances were taken in each as to indication or not of my university professor identity, for varied purposes of research facilitation and job acquisition and/or performance.

Conventional covert ethnographic research by a worker can provide the financial opportunity to earn while one learns, as a sort of research grant. It also can provide some personal gratifications that come from experiencing the challenges and satisfactions of a new or different work role.

In this type of research, the worker/researcher has a *special* ethical obligation. He/she must perform the work in as highly skillful a manner as is possible. This must be done in spite of work task adaptation difficulties and even if work imperatives must be given priority over research efforts.

This means that such research will usually take longer to complete than overt research by an outsider does. It will take more time to work and do research simultaneously, because such double-duty is not only difficult but sometimes even stressful. Double-duty note-taking and cautious conversational interviewing are sometimes helpful aids in this regard.

While overt in-depth interviewing is especially useful in identifying typologies and variations, covert participant observation excels in empathetic and emotional understanding of phenomena and processes. The latter also provides more "certain knowledge" about the occurrence of *in situ* events and definitions, as well as better insights into real/ideal behavior differences.

Finally, the conventional covert worker/researcher should try to "give something back" to the work or setting, in return for the advantage afforded in having been able to conduct covert research. One major way is through policy/practice recommendations in publications directly or indirectly related to the occupation or organization.

REFERENCES

Becker, H.S.

1951 "The Professional Dance Musician and His Audience." *American Journal of Sociology* 57 (September): 136-144.

1953 "Some Contingencies of the Professional Dance Musician's Career." *Human Organization* 12 (Spring): 22-26.

Bulmer, M.

1982 "The Merits and Demerits of Covert Participant Observation." Pp. 217-251 in M. Bulmer (ed.), *Social Research Ethics: An Examination of the Merits of Covert Participant Observation.* New York: Holmes and Meier.

Caudill, W.F., Redlich, F.C., Gilmore, H.G., and Brody, E.

1952 "Social Structure and Interaction Processes on a Psychiatric Ward." *American Journal of Orthopsychiatry* 22 (April): 314-334.

Dalton, M.

1959 *Men Who Manage: Fusion and Feeling in Theory in Administration.* New York: Wiley.

Davis, F.

1959 "The Cabdriver and His Fare: Facets of a Fleeting Relationship." *American Journal of Sociology* 65 (September): 158-165.

Erikson, K.T.

1967 "A Comment on Disguised Observation in Sociology." *Social Problems* 14 (Spring): 366-373.

Festinger, L., Riecken, H.W., and Schachter, S.

1956 *When Prophecy Fails.* New York: Harper and Row.

Friedman, N.L.

1967 "Career Stages and Organizational Role Decisions of Teachers in Two Public Junior Colleges." *Sociology of Education* 40 (Summer): 231-245.

1969 "Task Adaptation Patterns of New Teachers." *Improving College and University Teaching* 17 (Spring): 103-107.

1971 "Jewish or Professorial Identity?: The Priorization Process in Academic Situations." *Sociological Analysis* 32 (Fall): 149-157.

1973 "Orientations of Jewish Professors to the Jewish Community." *Jewish Social Studies* 35 (July/October): 90-108.
1983 "High School Substituting: Task Demands and Adaptations in Educational Work." *Urban Education* 18 (April): 114-126.
1987 "Reform Jewish Sunday School, Primary Grade Department: An Ethnography." *Jewish Education* 55 (Summer): 18-26.
1990 "The Hollywood Actor: Occupational Culture, Career, and Adaptation in a Buyers' Market Industry." In H.Z. Lopata (ed.), *Current Research on Occupations and Professions,* Vol. 5. Greenwich, CT: JAI Press.

Friedman, N.L. and Friedman, S.S.
1987 "Occupational Sociology as Career and Employment Counseling: Patterns and Possibilities." *The American Sociologist* 18 (Fall): 284-295.

Hilbert, R.A.
1980 "Covert Participant Observation: On Its Nature and Practice." *Urban Life* 9 (April): 51-78.

Holdaway, S.
1982 "'An Inside Job': A Case Study of Covert Research on the Police." In M. Bulmer (ed.), *Social Research Ethics: An Examination of the Merits of Covert Participant Observation.* New York: Holmes and Meier.

Humphreys, L.
1970 *Tearoom Trade: Impersonal Sex in Public Places.* Chicago: Aldine.

Lofland, J.C. and Lejeune, R.A.
1960 "Initial Interaction of Newcomers in Alcoholics Anonymous: A Field Experiment in Class Symbols and Socialization." *Social Problem* 8 (Fall): 102-111.

Mann, B.J.
1976 "The Ethics of Fieldwork in an Urban Bar." Pp. 95-109 in M.A. Rynkiewich and J.P. Spradely (eds.), *Ethics in Anthropology: Dilemmas in Fieldwork.* New York: Wiley.

Rosenhan, D.L.
1973 "On Being Sane in Insane Places." *Science* 179 (January 19): 250-258.

Roth, J.A.
1962 "Comments on 'Secret Observation.'" *Social Problems* 9 (Winter): 283-284.

Sullivan, M.A., Queen, S.A., and Patrick, R.A., Jr.
1958 "Participant Observation as Employed in the Study of a Military Training Program." *American Sociological Review* 23 (Decmeber): 660-667.

Whyte, W.F.
1984 *Learning from the Field: A Guide from Experience.* Beverly Hills, CA: Sage.

IMMERSED, AMORPHOUS, AND EPISODIC FIELDWORK:

THEORY AND POLICY IN THREE CONTRASTING CONTEXTS

Virginia Olesen

To re-review one's fieldwork years later is to undertake a geological dig wherein one partially uncovers strata of the sedimented self-at-work for current inspection and analysis, sometimes with insights not available at the time of the original study. One's own personal history, the reception accorded the work, changes in the topic or field studied all transform self and the original study as new relevances emerge and old ones fade (Nelson 1988). In recalling three studies in which I have seen myself as field-worker alter and change in the past 25 years, I have seen and felt my work change before the eyes of memory; indeed, at one capricious moment, I was tempted to entitle the chapter, "Remembrance of Things Asked."

Studies in Qualitative Methodology, Volume 2, pages 205-232.

ISBN: 1-55938-023-3

As I responded to the editor's invitation to recall the place of theory and policy in this work, I realized that the positioning of theory and policy in field work look quite different now than they did earlier. The abundant, even overwhelming array of older and recent thoughtful writing that links fieldwork to theoretical issues and vice versa (Glaser and Strauss 1967; Phillipson 1972; Scholte 1974; Glaser 1978; Halfpenny 1979; Spradely 1979; Emerson 1983; Hammersley and Atkinson 1983; Stanley and Wise 1983; Burgess 1984; Lofland and Lofland 1984; Silverman 1985; Manning 1987) sharpened my reflexive acuity, provided a background against which recollection could occur, but also tended to guide, even direct reflection. I have tried to strike a balance between what I could recall unaided and what this literature has brought to mind or even shaped. Since I also teach fieldwork and attempt to keep up with the avalanche of current literature, my classwork also served to remind and guide.

Literature that discusses policy and its place in ethnography or participant observation has also increased over these decades (Parlett and Dearden 1977; Cook and Reichardt 1979; Patton 1980; Bulmer 1986; Finch 1986). Again, teaching has both aided memory and framed it; several years ago with colleagues in medical sociology and medical anthropology on our campus we developed (to my knowledge) one of the first, if not the first course in the United States on fieldwork in evaluation and policy-oriented studies.

In this chapter I will reconstruct three studies which to date constitute my biography as a field-worker; they were not carefully selected to highlight different relationships of fieldwork to the times that the series editor selected. However, fortuitously, each of the studies represents a different configuration of temporal and contextual dimensions in fieldwork, dimensions that help frame the articulation of theoretical and policy-oriented issues in the work. The study of professional socialization, born out of policy concerns among nursing leaders and saturated from the outset with interactionist perspectives, took the familiar ethnographic form of continual, lengthy (three years) immersion in a single site. The research with temporary clerical employees, again generated by policy issues about urban life and informed by interactionist theory, was an amorphous fieldwork investigation where work site, though critical, was not directly accessible. Observation, the sine qua non of participant observation and ethnography had to be filtered through the workers' reconstructions of their experiences in various sites. Finally, the research on dissemination of an educational innovation (competence-based education) that was specifically designed to answer policy questions through an interactionist framework in my part of the study was centered on a site, but one that I visited only episodically as it was an hour and a half by air from my home in San Francisco.

As Hammersley and Atkinson (1983, p. 42) have cautioned, "It is important not to confuse the choice of a setting with the selection of a case for study."

Nevertheless temporal and contextual aspects of fieldwork frame methodological issues and strategies, points that I have developed elsewhere (Olesen 1982; Olesen 1983). These in turn link to the all important questions of whether findings relevant for theory and policy are credible and plausible. In the accounts to follow I will try to indicate in the reconstructions of my work where temporality and context shaped data gathering and analysis relevant for theory and policy.

LENGTHY IMMERSION: THE STUDY OF PROFESSIONAL SOCIALIZATION

As I look back to the 1950s and 1960s in the United States, those seem as halcyon days for field-workers: relatively easy access to health care and educational sites; available, even abundant funding; receptivity to publishing interpretative work (even though journals like *Symbolic Interaction* were far in the future); no institutional review boards that would later make certifying qualitative protocols so difficult (Olesen 1979a). (For non-U.S. readers, institutional review boards are federally and state mandated boards that review research proposals to ascertain that ethical standards regarding informed consent, risks and benefits and confidentiality will be assured during the research.) Chicago interactionism, though scattered because of the change in the direction of the department at the University of Chicago in the late 1950s, was nevertheless visible and viable; some of its leading scholars had already turned to the study of health care sites and professions.

Such was the climate of the late 1950s when the Dean of the School of Nursing at the University of California in San Francisco, Dr. Helen Nahm, a nursing leader of national renown, determined to strengthen her school and to demonstrate the place of the social sciences in nursing education. To achieve this she recruited Anselm Strauss from Michael Reese Hospital in Chicago. Strauss was already well known for his interpretations of George Herbert Mead (1946) and his own classic, *Mirrors and Masks* (1959). Together they tapped the then abundant funds for research in nursing issues at the Nursing Resources Division of the U.S. Public Health Service. They framed a study which Nahm believed would be significant in understanding the processes through which baccalaureate students of nursing acquired a professional identity and an orientation to nursing leadership. (For non-U.S. readers, nursing education in the United States has been in the university since 1909, though it also occurs in hospital schools, community colleges, state colleges and private colleges and universities.)

Once the grant came in, Strauss activated old Chicago networks to recruit the team: Fred Davis, who had worked with Strauss at Chicago, came on as project director; I learned of the project from David Riesman, whom I had known while in a master's program in mass communication at Chicago and with whom I had stayed in touch when I went on to Stanford; Elvi Whittaker, one of Kaspar Naegele's students who had done a study of student nurses, was moving to the Bay Area. So the team was complete.

My own credentials as a symbolic interactionist were only minimally viable. Although I had vivid recollections of Strauss' memorable class in social psychology at Chicago, I had not extensively used the interactionist framework in either my master's thesis or my doctoral work at Chicago but I had remained intellectually sympathetic. I was somewhat better equipped to do fieldwork, for prior to graduate study I had had a long (seven-year) career as a reporter and editor, where I had acquired some journalist's skills of observation and analysis that resemble, if not replicate on some occasions, those of the fieldworker.

As the research got underway at San Francisco, my earlier work experience in journalism became integrated with the vibrant encounters in sociology in Chicago and the introduction at Stanford to the then new field of medical sociology and the old field of cultural anthropology. Doing fieldwork on site (the school and the health science campus) and analyzing data with Fred Davis (1979) and Elvi Whittaker (1986), both of whom were and are gifted analysts of qualitative data, facilitated this integration and my own transformation from a journalist-sociologist looking for "facts" to a conceptually oriented fieldworker in the interactionist tradition. Busy with his colleagues Jeanne Quint Benoliel and Barney Glaser on his own studies, those of dying patients, Strauss did not participate in our research, but frequently contributed ideas and suggestions at our team meetings.

Theory in This Study

What part did theory play in this study? As the extensive details recounted elsewhere about our methods and our analysis show, we borrowed early and heavily from symbolic interaction (Davis 1968, 1972; Olesen and Whittaker 1986). How we saw students interact with multiple others, fellow students, faculty, patients, boyfriends, staff nurses, physicians, and very importantly, ourselves and how they spoke of these interactions to us and to others were empirical issues given in the theoretical framework of symbolic interaction, which is ideally suited to analyses of process. In fact these themes led us to find the work of Merton and his colleagues (1957) on the socialization of the student physician somewhat wanting, for that work did not touch on the multiple roles in the socialization process, roles that included not only student and student nurse, but adult woman, and so forth. More useful for us because it fit the interactionist framework and assumed an image of the student congenial to our thinking was *Boys in White* (Becker, Geer, Hughes, and Strauss 1961). Their analysis of how students create student culture and the restraining and guiding influences of norms within student culture accorded with our views that the student nurses were by no means passive creatures in their learning, but were active participants in their own socialization.

Theoretically, however, our work was more than merely derivative of *Boys in White*. Rather early we became highly attuned to phenomenological and existential sociology, particularly the work of Alfred Schutz, much of which could be integrated with the interactionist framework (Schutz 1962). This orientation, plus our utilization of Goffman's influential dramaturgical analysis, was continually present both as guide to the fieldwork and resource for analysis (Goffman 1959).

The interplay between the theoretical perspectives on the one hand and emergent events was critical for us. Naturally, we were following those personal and curricular events as they occurred. In this sense the content and pacing of the curriculum influenced the rhythm of our observations, but the interactionist and phenomenological theories enabled us to observe and analyze events and features of the socialization process in much wider terms than merely relying on "the impact of the curriculum." That would have obscured the active part students were playing in their own socialization and the many factors in that process. Moreover, as was consistent with our theoretical orientation, we could regard the curriculum as a set of symbols and meanings, which students and faculty continually altered and reinterpreted in different contexts, a view quite contrary to that of many curriculum planners in and out of nursing who tend to envision curriculum as stable, if not ossified. Our view of curriculum as a set of altering symbols sidestepped treating the curriculum as an "independent variable" or sole factor, a position that would have tied the study too closely to the San Francisco school. (For a review of our strategies to introduce comparative analyses, see Olesen and Whittaker 1968, pp. 24-25.)

A single issue, the place of emotion in professional socialization, will illustrate the interplay of theory and fieldwork in this study. Because we were keenly interested in the dramaturgical aspects of self and socialization, we did highly detailed observations of students' interactions with others, their language, their behavior and importantly, their verbalized and visualized expressions of emotions.

Participant observation, when linked to a theoretical framework such as symbolic interaction in a situation where observer and participant share cultural meanings, offers a powerful feature, the observer's opportunity to witness emotion, as well as to hear about it from the participants and to recognize in herself or himself if not the same emotion, a sympathetic response to it. Expression of emotions in fieldwork constitute supreme moments of reflexivity for the observer. Even as I write this, some 27 years later, I can vividly recall—without field notes, instances of students' display of emotions and my own responses. Indeed, when I teach fieldwork, I urge students to capture those emotional moments and to reflect on them once out of the field.

It was our observation of visible depression among the students, a mere six weeks into the study and the students' program, that compelled us to orient our analysis at the time and through the remainder of the study to the occurrence and expression of emotion among students and to the place of emotion in professional socialization. We had first noted that some students seemed "down," but as the days went by and students talked about their feelings with others, the pallid faces became more noticeable, the expressions of despair more frequent. Needless to say, we undertook intensive interviewing and observation to follow this developing and enlarging phenomenon and spent countless hours attempting to account for it within the interactionist framework: loss of self, communication of feelings, which in turn led to further questioning and observing (Davis and Olesen 1963). As more students declared themselves depressed, anxious and fearful for their future in nursing and as women, their depression was not only a manifestation of subjective assessment, the self talking about the self, but it became, we could see, transformed into collective expression and commonly understood norms: "There was much evidence to indicate that, while by no means every student felt the deepest of depressions, the depression itself became somewhat of a norm in student culture" (Olesen and Whittaker 1968, p. 250).

At the time of the depression our analysis that related the phenomenon to interactionist theory was preliminary, for the fully developed analysis, utilizing the concepts of status transition and identity stress, both drawn from interactionist frameworks, did not occur until nearly the end of the study when we had more time away from the field. The event, however, and our preliminary analysis of it, sensitized us both to the observance of emotion, its occurrence and expression, and some of the then current interactionist work on emotion (Goffman 1956; Gross and Stone 1964; Lynd 1958); as well as theories of Simmel (1950) and Scheler (1961). Later, after we left the field, when we analyzed laughter among the students we looked at psychological and phenomenological literature on emotions and integrated the psychological conceptualization of release of tension, expressed in laughter, with an interactionist perspective on a situation when a student would make an "identity error" and create a tense situation for herself, the other students, the instructor, and ourselves, the observers (Olesen and Whittaker 1966).

Returning to the interplay of theory, observation and analysis, which were intertwined from the first day and became increasingly so as we carried the work forward, the phenomenon of the depression had sensitized us to the *occurrence and experience* of emotion in the students and the possible interpretations of that for socialization theory and for further observations. Interactionist theory at that time had not developed the issue of emotional occurrence to the extent that is now the case; emotions were still the province of the psychologist. Since we were not working with Freudian interpretations of the socialization process, we assumed as had been the case with the collective depression that emotions were indicators of subjective aspects of socialization, as regards an emergent self and professional identity but were not personality or characterological issues. Scattered throughout notes, though not fully analyzed until much after the end of the fieldwork, were discussions of what these occurrences signaled for the loss of self or the anticipation of a new view of self. Later when Elvi Whittaker and I struggled with conceptualizing the dynamics of socialization, we analyzed the minute interaction and phenomenological processes, borrowing Hughes' concept of "cycles and turning points" (1984a), McCall and Simmons' "legitimation agenda" (1966, p. 165), Strauss' "status forcing" (1959, pp. 84-87). The occurrence of emotion, we argued, provided part of the dynamics of these cycles, which we termed adjudication and legitimation of the emergent self (Olesen and Whittaker 1968, pp. 223-232).

We were, however, also keenly interested in a different theoretical angle on emotion, namely the *expression* of emotion and the circumstances, for example, displays of sympathy, irritation, joy, and how those expressions were socialized in interactions with each other, the faculty physicians, patients, and ourselves. Here the issue was one of presentation and what that presentation suggested about the self being presented. As the years passed our field notes filled with observations of the students' early, awkward emotional expressions with patients and house staff on through the smoother, "more professional" display of a student who was becoming "a nurse." However, this dramaturgical angle on emotion in our work became somewhat submerged in a concept that we developed to designate how students learned to present themselves to various others, namely, the concept of "fronting" (Olesen and Whittaker 1968, pp. 173-184).

Had some of the much later work on the social psychology of emotions been underway, work that is rooted in symbolic interaction, we might well have traced more finely the rules for emotional expression and how students were socialized to those rules (Lofland 1979; Charmaz 1980; Scheff 1983; Denzin 1984; Hochschild 1984). At one point we considered doing an analysis of how different professions socialize students to emotional understanding and display of emotions, but took this idea no further than a brief excursion into descriptions of students studying law, divinity and medicine in order to try to provide comparative materials for our own findings in professional socialization. The surge of writing, research and theoretical exploration of the sociology of emotion of the 1970s and 1980s was aborning as we did our work, but contributions from that genre lay far in the future. Our own work remained closely tied to the theoretical issues in symbolic interaction and the dynamics of professional socialization, not the sociology of emotion. The analyses of emotion in our work widened interactionist perspectives on professional socialization, adding new elements to the consideration of that process.

To return to the dimensions of context and temporality in the fieldwork which I mentioned in my introduction, I believe that we were able to pursue the dialectic of theory and observation with emotional expression and occurrence and to generate theoretically relevant findings because we were immersed in a single site for a lengthy period. (This point applies to many of our other findings as well.) That lengthy immersion provided us the opportunity to observe, think, re-observe, re-think, and re-analyze. Above all, it enabled us to *feel and understand* students' experiences and emotional expressions because, to use Schutz' apt phrase, "we had grown older together" (Schutz 1962, pp. 220). Thus the field-workers' reflexivity in these kinds of analyses adds to the power of the method, but the temporal and contextual dimensions of the method facilitates such reflexivity, particularly when we have what Foster has termed "staying power in the field" (Foster, Colson, Scudder, and Kemper 1976).

Policy Issues in the Socialization Study

The theoretical framework of symbolic interaction allowed us to address the key policy issues of interest to nursing leaders that had led to funding for the study: commitment, career and leadership potential. These widely debated issues of that time touched curriculum planning in university schools, as well as directing the orientation of major nursing organiztions and the federal funding of training programs for nursing education and research.

Cast into the conceptual dimensions of symbolic interaction, these policy questions guided fieldwork with the students and framed our use of concurrent methods, which included observations and interviews with several classes of students and periodic questionnaires for all classes enrolled in the school during the three years of fieldwork. These multiple methods introduced a comparative perspective into our work and provided a stronger base for the theoretically oriented and policy-related findings than would have otherwise been the case.

Policy, framed in the concepts of symbolic interaction, had outlined much of our study. Those issues, unlike the theoretical questions, did not lead to the same interplay between fieldwork and questions, event though we were mindful of them. Unlike field-workers who present policy-related findings during the course of fieldwork to those with whom they have been doing research, we waited until the end of data gathering in 1963. We were concerned that early feedback at San Francisco would alter "the natural course of events" and perhaps even damage our intimate relationship with the students. It did not seem appropriate to discuss the study elsewhere with nursing leaders.

Once the fieldwork had concluded and we faced the monumental analytic task of organizing our fieldwork notes and the quantitative materials, we began to communicate our findings to nursing educators and others via various consultations and speeches. We also began to publish extensively.

Of the work that came to print, one paper dealt directly with policy issues and made specific suggestions for policymakers (Davis, Olesen, Whittaker

1966). In that paper we had found it relatively easy to link the theoretical issues with policy questions. Because we had early articulated a conceptualization of lateral roles and selves undergoing socialization (e.g., nursing, women, adult, friend, and so forth), which was theoretically innovative at that time, we could discuss students' future commitments and careers not merely in terms of future roles as nurses, but as women. Though the findings were qualified, this latter role and image of self cast in a conventional female frame, seemed to us to outweigh the other roles and selves: "the solution to the career commitment problem resides much more with American culture at large, particularly with the mores governing adult sex roles, than it does with the professional as such" (Davis, Olesen, Whittaker 1966, p. 174). Relying on our analyses of the degree to which the students incorporated the imagery of the public health nurse into their self concepts and were, therefore, reluctant to envision themselves as hospital nurses, we could also point out that this particular baccalaureate education, with its emphasis on public health values, had succeeded only too well. This had potential policy implications, since most of nursing work did and still does occur in hospitals.

Many of our other publications presented highly phenomenological analyses and findings that were of interest to sociological social psychologists and nurses trained in that discipline, but they did not present specific policy suggestions. Many contained findings that did bear on policies for curriculum and faculty development in university nursing schools, for instance, the contribution of student culture to professional socialization. Whether these findings were influential for or transformed into policy is a difficult question to answer, for concerns in nursing shortly after our work was over (1967) moved away from focus on students and shifted to clinical practice in various settings, especially the hospital. It is my own impression that those findings were and are of greater interest to social scientists and research nurses in the United States and Britain than they were to policymakers in American nursing, whatever the level.

The wide dissemination of our findings, though perhaps not widely assimilated in nursing policy, did, however, in my view, serve to establish, along with the work of Strauss and his colleagues, fieldwork as a legitimate research method in nursing research that had long been dominated by positivism through the tools of educational psychology. Thus the findings generated in a study originally oriented by educational policy became the vehicle through which nursing researchers could borrow and adopt a method new and to them innovative and in so doing create policy issues in curriculum around nursing research that are still, 20 years after our work, lively and even controversial.

Postscript

Our field work experience was somewhat unusual in that we did not leave the field in 1963 when we ended participant observation. We remained on the site as researchers until 1967, but were no longer involved in studying students, but rather with the analysis of our work.

Subsequently we could and did turn our attention to our former participants, no longer students, but graduates. With wisdom born of hindsight we congratulated ourselves on having utilized—along with our fieldwork—periodic questionnaires while the study was underway. Now we could follow up the graduates, even though we had stayed and they had scattered. Two follow-up studies in 1967, 1974 used interviews and questionnaires, while the 1985 follow-up, a pilot study, relied on interviews. Thus our fieldwork has been transformed into a longitudinal analysis, but with the same theoretical orientation and similar questions regarding career, commitment and leadership, since professional socialization clearly does not end at the door of the university (Olesen 1973; Lewin and Damrwell 1978; Lewin and Olesen 1980; Olesen 1987).

AN "AMORPHOUS" FIELD: THE STUDY OF TEMPORARY CLERICAL WORKERS

In the early 1970s, weary of the socialization study and looking for fresh research challenges, I exhumed an old memo I had written to myself in 1968 on the possibilities of doing an urban ethnography with temporary clerical workers. (In Britain Alfred Marks and other similar agencies hire and place these workers; in the United States they at one time were known as "Kelly girls" after the name of an older firm in this field. In the United States they are now termed "temps.") Therein I had asked myself a research question that reflected some of the theoretical interests of the socialization study: how does the temporary employee create and sustain an identity as a worker if she/he is continually transitory? My interest partially grew out of my own graduate student experiences at the University of Chicago in the mid-1950s. Before I became aware of the world of grants, fellowships and stipends, I had supported my master's studies by working in the clerical pool at Billings Hospital, the University's teaching hospital.

My research question also emerged from my criticisms of sociological and feminist scholars on occupations, professions and women at work. At that time, before the recent surge of writing about housework, waged work, clerical workers, I believed the literature to be overly oriented to elite occupational pursuits, women doctors, for instance, or to exotic or deviant work, blue-collar workers, prostitutes. These studies overlooked occupations where most American women who go into the labor market find employment, namely, clerical work. I was also intrigued with how these workers fit into urban life and wrote in that memo: "These urban nomads strike my researcher's fancy; study of their working lives seems to present not only an opportunity to understand a little understood sector of our urban scene, the clerical world, but also suggests a context in which the intriguing conceptual issue of alientation in work seems paramount" (Olesen 1968).

At that time federal research policy still took as a critical topic the issue of urban problems. In the National Institutes of Mental Health, Eliot Liebow, a field-worker of note for his ethnography, *Tally's Corner* (1967), headed a Center for the Study of Metropolitan Problems. It seemed a likely place to propose the research I had in mind, for it offered what was then known as "small grant support" (up to $5,000) which meant that a modest beginning could be undertaken and if all went well, I could step to a larger study. (In those days $5,000 was a very small grant; in today's barren funding climate for qualitative projects, it seems handsome, indeed.)

Happily, my hunch was correct; the small grant was funded for a year. After some floundering, with the help of the president of the California Association of Temporary Services firms, I secured the cooperation of four Bay Area companies. Almost immediately problems of access, ethics and theory converged to shape what I would be able to do.

Temporary services employees register with firms who specialize in such placement, take a test to see what skills they have, and then phone in to an employment "counselor" who gives the temporary worker an assignment that is presumably commensurate with the individual's ability and which may be for a few days or as long as a month. It became apparent that we would not be able to accompany workers on their assignments and to observe them and others in the work site. The task of explaining our study to innumerable office managers and other workers plus the psychological and social burdens of hanging around with the workers appeared ethically and strategically difficult. We did not have the luxury of untrammeled time in which we could have waited for the occasional chance to go out with an employee; the grant was small, the time limited to a year. Moreover, the graduate assistant had a heavy schedule of classes and my own life was quite hectic.

There was a "field" there to be sure, but one in which we could not physically participate and observe as we had with the students in the socialization study. There students learned to manage our burdensome research presence and our long-term immersion meant that we could reciprocate in an interpersonal way for the cost of those burdens, something that appeared quite impossible with the temporary employees. Further, neither Fran Katsuranis nor I typed well enough to fake becoming temporary clerical workers, a charade that would have been enormously time consuming and potentially unethical in my view. (By this time we had an Institutional Review Board on our campus, and as a consequence of my criticisms regarding its overly rigorous review of qualitative research at UCSF I had been placed on the Board. Even if I had not already regarded this strategy as unethical, I would have had to confront review of my own work by this same board that surely would have found the strategy wanting in terms of disclosure, and so forth).

I came to the conclusion that in this "amorphous" field that detailed interviewing with employees, the employment counselors and the owners of the firms that "the field," that is, the work sites, would have to be re-created in those accounts. The "field," of course, included the telephone contacts with the supervisors, which in any instance would have to be reconstructed. From the standpoint of the theoretical issues of initial interest in the study, identity construction and alientation, heavy reliance on careful interviewing would be critical in order to evoke the details of the work settings and the individuals' experiences in those, as well as their views of themselves. As this strategy emerged, I thanked my lucky stars that I was working with Fran Katsuranis, a former psychiatric nurse who was a superbly skilled interviewer. Her interviews with the temporary clerical workers drew out detailed accounts of their work experiences, their lives and views of themselves. Fortunately, the study itself appealed to the workers we interviewed, for we had little trouble eliciting extensive detail.

As I now recall our analysis from re-reading some of our analytic memos done during the course of the work, the theoretical interest in how temporary workers constructed identity led us to re-think the question of alienation in their work (Blauner 1964). Rather early, even as we were still interviewing, we asked detailed questions about the workers' conversations with their employment counsellors. These questions were essentially influenced by an interactionist paper on the sociology of telephone conversations (Ball 1968) but focused on the issue of how workers presented themselves in these conversations in order to achieve good assignments. Just exactly when the question emerged of "who has the upper hand," or phrased more elegantly, who controls whom in these types of exchanges is now unclear to me, but it was not long before we realized that the workers were not mere pawns in the hands of the firms. Both they and the placement counselors reported that many had worked out for themselves ways to negotiate better jobs and better assignments, often full well realizing what one counselor told us: "We can get customers more easily than we can get girls." Some workers even reported deliberate strategies to manipulate supervisors that helped keep the worker visible and knowable and above all, defined as a reliable, highly skilled person who would do well for the client.

Clearly, we had to think about what this meant for our interests in identity construction. Workers had to present themselves via the phone as skilled and ready for assignment, but unwilling to take just any assignment—within limits. Too much bargaining or too many refusals would lead the supervisors to offer the work to someone else. On the other hand the workers told us they also knew that there were times when the firm had more assignments than workers, hence they had the "upper hand."

What we began to realize as we thought this issue through was that the question of alienation related not only to the work site, where indeed, many workers reported difficult or unpleasant situations. Since the telephone encounters were in fact a part of the work, obtaining the assignment, there was, by all accounts, considerable worker control over refusal or acceptance of an assignment. While it was true they had little or no control on the site, they could and did exercise control in obtaining the assignment. Since alienation in work is customarily thought to relate to, among other elements, control of work, here was a phrase in the work cycle where these workers clearly had control, at least to a degree recognized by themselves and the firms for which they worked. As we wrote in our final report: "In short, the simple matter of worker refusal becomes a type of control. It may articulate the sort of job assignment the individual gets, but perhaps more importantly, it gives the individual a sense of self, a limited sense of control over one's fate and destiny in a very ambiguous and isolated type of work experience" (Olesen 1974b).

Though we could document that they were exercising control of a sort in the conservations with the placement supervisors, we still had to unpack the question of the rest of the work experience. How did they see themselves in the work site? What were the interactions there that influenced views of themselves as workers? Again, our interviewing had to probe deeply for descriptions and discussions of work site experiences, pleasant and unpleasant; relationships with others, and above all, how they defined themselves as workers and as temporary clerical workers. The constraint of lack of access to the actual work sites coupled with the theoretical importance of this question to the study pushed us into deeper interviewing than might have been the case had we been able to observe.

As we analyze their accounts on these issues, we could see that they were revealing themselves to us in much more complex ways than our original theoretical position had suggested, for example, that work could be a "master" status (Hughes 1984b, pp. 141-150). Clearly, we had to move to a broader formulation and we invoked the conceptualization of multiple roles and selves, or work and lateral life roles that had been so productive in the earlier socialization study. Widening our theoretical scope enabled us to conceptualize the complexities in their constructions of self and to see more clearly the place of temporary work in these complexities. The place of temporary work as identity resource had to be analyzed in a more complex frame, for instance in order to understand those who saw themselves as temporaries but in connection with another view of self that was often more important, for example, craftsperson, weaver, musician (Olesen and Katsuranis 1978). I believe this analysis became as refined as it did with regard to control in their work and construction of identities because we had to interview in depth and then to analyze with great care in lieu of being able to observe. Thus, theory, methodological strategies and analytic pursuits were integrated in a study where the "field" was "amorphous."

Temporary Employees' Study and Policy

The theoretical issues in temporary work were much more central than were policy questions, but the orientation of the Center for Studies of Metropolitan Problems necessitated that we remain alert to some policy implications of the study. The Center's liberal outlook engendered requests to grant holders to present to audiences where the research findings could be transmitted to individuals who might apply the findings or find them useful for policy purposes. In 1974 we moved away from the theoretical perspectives of worker control and construction of identity to look at the industry itself for a meeting on "The Urban Struggle" at the American Psychiatric Association. At that panel, which the Center had organized, we argued that temporary services firms, because of the industry's ideology that everyone can find some work through temporary placement, actually helps integrate workers whose chances for steady or permanent employment is minimized by problems with age (very young or very old workers), substance abuse (alcoholism or drugs), lack of training or outdated training.

In this presentation we had to place our contextualized fieldwork in an analytic frame which spoke to the place of work and "deviant" workers in American society: "In an industralized bureaucratized society which prizes identity based on productive work, yet which denies work to certain persons, temporary services firms redistribute and reintegrate persons who are otherwise marginal to the labor force and are sometimes unemployable on a steady basis. Whether this is the best, if there is a best, way to redistribute and integrate such persons and whether it provides a sound basis for occupational identity in the urban struggle cannot be answered here. What our data from the firms and the employees suggest are the integrative functions of one sector of Amerian society in which social scientists and psychiatrists customarily do not expect such integration" (Olesen, 1974a).

Aside from this presentation and others made to applied social science societies, such as the Society for Applied Anthropology, we circulated our work to the firms and their organizations. We thought there were some clear policy implications in our work; the findings might well have framed guidelines for reception and relationship with temporary workers and for some understanding of the meaning of the work in their lives. We did not present to the national or regional meetings of the temporary services firms, for the political economy of energy in a university which demands documented achievement in scholarly modes simply would not stretch to include presentations in these non-academic areas.

I believe this raises the question of demands on the field-worker's time and locale as policy issues. Theoretically oriented research even with implications for policy may not be easily disseminated when the field-worker's locale is quite different from that where policies are made. Had we worked out of a business school, or a policy program we might have more readily found avenues for presentation. In retrospect I also think that had some of the organizations devoted to well being of clerical workers existed then (e.g., Nine to Five), we could have disseminated our findings through them. Our findings did filter into other social science publications on clerical employees (McNally 1979) one of which spoke directly to the policy issues that emerged later around electronic technology and clerical employees (Lewin and Olesen 1985).

EPISODIC FIELD WORK: STUDYING AN INNOVATION

By 1974 we were winding down the study of temporary clerical workers. However, I was not rushing to start new work or even to amplify that study. My university commitments had expanded to chairing our then new Department of Social and Behavioral Sciences, I was busy organizing the first national conference in the United States on research on women, health and healing and there was serious illness in my family. Yet when Gerald Grant, a leading scholar in the sociology of education and formerly education editor of *The Washington Post,* invited me to participate in a national fieldwork study, the prospect of new challenges in fieldwork was too tempting. Grant was putting together a team to study the dissemination of a type of educational ideology, competence-based education, which was then spreading widely in American higher education. I knew little about the sociology of higher education and in fact had rather avoided it up to that point, assuming rather naively that the drearier aspects of curriculum planning and analysis of college students' grades prevailed, an assumption that I quickly dropped as I became aware of the sophisticated work done by Grant and others who were joining the team, David Riesman and Zelda Gamson (Riesman, Gusfield, and Gamson 1970; Riesman and Grant 1978). I did have long-standing interests in the dissemination of innovations plus the research experience in the earlier study of baccalaureate education in nursing. So, in spite of my other responsibilities, I agreed to participate and to conduct the fieldwork at one of the twelve sites to be studied, the nursing program at Mt. Hood Community College near Portland, Oregon, an hour by air from San Francisco.

My hunch that this would prove intellectually and professionally stimulating was correct. Our team included sociologists, a philosopher/psychologist, an historian, a professor of English literature and creative writing, and several of Grant's graduate students. In sum, a truly interdisciplinary undertaking. The team meetings constituted a veritable Athenian marketplace wherein stimulating ideas, discourse and argumentation abounded, demonstrating to me that individuals from disciplines other than sociology and anthropology can do insightful fieldwork and analysis. One could not be too stuffy about sociological theory in the face of witty and incisive comments from the philosopher or the historian or the professor of English. Throughout my career as a field-worker I have been lucky to have worked with gifted individuals from whom I have always learned a great deal; my experiences with Grant's team continued that good fortune.

The design of the study was such that we were to do at least twelve days of fieldwork per year on our own educational site and additionally, participate in the quarterly team meetings held on different sites. There we did comparative observing at that site and additionally discussed one another's field notes that had been shared since the last meeting. This provided an invaluable aid to checking our own interpretations and gaining ideas for new leads. The very idea of doing fieldwork in this temporal frame, an episodic study in a single site, looked challenging, for it meant that my fieldwork approaches, learned in the leisurely years of the socialization study would have to be refined or even modified.

Because Grant's account of our methodology is extremely detailed, I will not repeat those details here, but reconstruct my part of the study to highlight my own theoretical itnerests and how those dovetailed with the project's policy orientation (Grant 1979a; Olesen 1979b). For me the study posed the chance to do a theoretically-oriented study that was much more closely integrated with policy concerns than either of my two previous studies had been. Because the project was oriented to how competence-based education was being implemented around the United States, a problem I construed as a dissemination issue involving process, I began shaping my fieldwork once again around the theoretical dimensions of symbolic interaction, which seemed most suitable to the problem at hand. Most of the sociologists on the team were cordial to, if not actually highly conversant with interpretive sociology, including symbolic interaction. The others, of course, did not share this theoretical perspective, but being humanists, took the world as continually construed and reconstructed by meaningful human action informed by openness and choice, a viewpoint certainly congenial to the interactionist's outlook.

Competence-based education refers to a complex movement in American higher education that moved through the United States in the late 1950s and early 1960s. This type of education specifies outcomes for students and demands assessment of those outcomes. Student learning is conceptualized in terms of these outcomes and progress is certified in terms of achievement with regard to those outcomes. Though it clearly had origins in the early measurement of behavioral objectives in higher education, it began to assume reformist overtones when some educators saw it as an opportunity to enhance learning for disadvantaged minorities, as well as to assure that graduates could *do* something; "A number of colleges—primarily the nonelite sector of American higher education—have recently attempted to reconceptualize what it is they do, with the aim of being able to state that their students are competent *at* something or competent to *do* something rather than that they have accumulated so many course credits. This is the heart of competence-based educational reforms" (Grant 1979b, p. 2).

Some of the colleges referred to in the quotation had received federal funds from the Fund for the Improvement of Post-Secondary Education, Department of Education, to implement competence-based programs, grant[5] made within the reformist orientation of the Fund. (The Fund still operates much like a foundation within the federal government, though the reform orientation differs with the philosophy of the Secretary of Education and the current administration.) In 1973 FIPSE (the acronym for the Fund) funded a three-year research and evaluation project to assess how these and other programs "have been conceptualized and operationalized, to discover how these settings have been created and evolve . . . to describe the programs, probe their assumptions, explore the pedagogic issues they raise and in an open-ended way assess the range of impacts they have" (funding proposal quoted in Grant 1979a, p. 443). In accepting the leadership of this complex project Grant, who had been influenced by the interpretive approaches to fieldwork and evaluation and particularly by Parlett and Deardon's "illuminative evaluation" (1977), conceptualized the research itself as an experiment. Rather than digging out enumerative data on which so many evaluation studies depend (how many indicators show how much), this study, as envisioned by Grant, would capitalize on the emergent or naturalistic properties in classical ethnography: "Our research project was an experiment in the sense that it posed a test of the question: what is the best way to study an emergent large-scale social movement or educational reform in order to produce the greatest yield of useful knowledge to policymakers and potential participants in such a movement or reform—not merely descriptive information about particular programs, but knowledge about what it would mean to foster more widespread adoption of the reform?" (Grant 1979a, p. 440).

Grant's creative conceptualization of this research deflates criticisms that ethnography or fieldwork methodology cannot be used to study phenomena on a large scale, particularly when resources, both intellectual and financial, of this magnitude are available.

Not only was the design of the study complex, but there were several levels of relationships that intertwined at numerous points. There were, of course, the interactions and fieldwork relationships at the site. Working at top speed, though not without difficulties that I have outlined more fully elsewhere (Olesen, 1979b), I established and maintained more or less successfully—relationships with administators, students, faculty, Oregon nursing leaders. This proved to be far more difficult than I had anticipated, for mid-way through the study, students rebelled against the competence-based program and obtained the support of the administration that forced out the nursing faculty that had started the innovative program, and I had to start anew.

Then the research team itself posed another level in establishing and maintaining relationships. In such distinguished company I certainly did not want to be seen as difficult, dense, uncooperative, and above all, a weak field-worker, especially when I had been bid to the team on the basis of my past ethnographic work and when my university department was widely regarded as a center of fieldwork. Though relationship within the team were with very few exceptions unfailingly cordial and supportive, and, as I have already indicated, wonderfully stimulating, nevertheless I always felt that I had to be "up" for the team meetings, particularly when we discussed wide-scale policy issues in higher education, a scope well beyond that in my previous work.

Finally, there was the relationship of the research team to the funding sponsor, FIPSE, with the Fund's expectations for answers to what they saw as critical dimensions in this educational reform. The Fund itself was not without complex problems, for, being beholden to the Congress for support, the directors were anxious that all the Fund's project produce "something useful."

In this third level of relationships, obviously integrated with the other two, intense conflict between the team and the Fund arose over our roles and researchers and evaluators. More specifically, as Grant has lucidly outlined in his methodological account of our work, the Fund wanted to know, among other issues, what "the outcomes" would be (Grant 1979a) and how feedback was to occur to the sites where the work was occurring. This issue, of course, invoked the always present problem of confidentiality at the sites, as well as the nature of our roles as researchers. To what extent were we evaluators and to what extent field-workers? The Fund clearly saw us as the former; most of us, myself included, regarded ourselves as the latter. The nub of the tensions, later resolved by a mass meeting between the research team and project directors at the various sites, however, lay in the divergence between our research orientation and the reform outlooks of the Fund. Grant has captured this well: "Our first surprise resulted from our failure to see how starkly the logic of our research methods contradicted the logic generating the reform we studied. As mentioned earlier, we proceeded by inductive and exploratory methods, discovering, testing, and modifying hypotheses as we enlarged our understanding, whereas most persons at the sites in the study proceeded deductively, deriving curriculums from highly specific sets of desired outcomes" (Grant 1979a, p. 486).

As for my own part in this controversy over policy between the research team and the FIPSE staff, when Grant had invited me to join the research and explained the nature of the study, I knew immediately I would be unable to provide data that would "evaluate" the Mt. Hood program as a policy topic. My experiences in the San Francisco study of student nurses made me aware of how complex curriculum implementation is and how highly varied student responses to it can be. I was also mindful of the complex politics in nursing education and I knew very well that curriculum issues, even in sociology, did not excite me, though I recognized their importance. What did intrigue me was the question of how a set of ideas became defined, accepted, rejected, modified or mystified and by whom in the community college context. Though I did not realize it at the time, these *were* policy issues. Not until I read some years later a classic paper applying the perspectives of symbolic interaction to policy analysis in which policy as "the transformation of intentions" was discussed did I realize that my analysis indeed was policy oriented, though not in an evaluative way (Estes and Edmonds 1981).

If theory and policy were closely integrated in this study, the mode of fieldwork, episodic in a single site, made pursuit of the understanding of these quite problematic. The episodic nature intensified the relationship among and between typical problems that every field-worker encounters and the theoretical and policy issues.

In a site where I had been continually immersed as was the case at San Francisco, the temporal abundance allowed opportunities to establish relationships, continually work at them and the data they produce. Those were not easy when I was on site several days every few months. (Recall that we were to spend at least 12 days a year on site; I spaced these so as to visit periodically at the end of school terms to see what had happened in that term.) Because of this I found entree very difficult: the Mt. Hood faculty harbored ambitions to do their own book on competence-based education in nursing and were wary about an outsider, particularly someone who might be a spy for FIPSE that had sponsored their program. (This was yet another reason why I resisted defining my work as "evaluation" as the FIPSE staff wanted us to do.) Moreover, the director of the program was initially completely inaccessible. Later I learned she was terminally ill with cancer, a fact which had been concealed not only from me, but from others at the college in the way that sometimes happens in such tragic situations.

What these methodological difficulties meant for theory was that on the spot data in every encounter had to be carefully assessed with regard to the quality of the data (were people lying or covering up?) and analyzed as to how it fit with my hunches, leads, propositions. This, of course, happens to a certain degree in all fieldwork studies, but here every moment had a "data weight" attached to it. As I wrote to myself in a memo about this type of fieldwork: "If I think I have to be sensitive in the ordinary fieldwork situation, this work poses another order of magnitude far beyond that. I have to watch every cue, the verbal, the nonverbal and what happens at a particular point in time, the kinds of illusions and allusions, the very structure of their language. It's like being a safecracker; I've got my mental fingertips so finely sandpapered that when I 'start to open the safe,' I can almost feel and hear the tumblers clicking inside the heads of my respondents" (Olesen 1977).

Because I was interested in competing definitions of the competence-based progam, I had to be sure I left no one out who might possibly guide, influence or contribute to shaping the definitions, even by lack of interest in or hostility to the program. Whereas in the socialization study, we could concentrate on the students, at Mt. Hood in the span of several days I would have to spend time with first- and second-year students, faculty, administrators, registrars, nursing leaders and graduates of the program. I could not hang around the student lounge drinking coffee and chatting with students for long periods of time. I had to plan for and think through each encounter, but I also had to allow for enough flexibility in the visit to focus on theoretically relevant themes that would have emerged since the previous visit or during the visit itself. Some would perhaps criticize this as not being "naturalistic," but in point of fact to a lesser degree we did this in the other two studies about which I have written, and in other studies where one is continually immersed, as reading any ethnography makes clear. Though one constantly remains alive to the "natural" flow of events, sharpening hunches, developing leads, checking older themes, looking for negative cases all guide the field-worker's day. One does not bob about directionless like a cork on the ocean. This in any case would be impossible since as an interacting, reflecting actor in the situation one's very presence partially guide events.

Added to these temporal exigencies were problems around the fieldwork that developed at Alverno College, another site in our study where there was an important and influential program in competence-based education, one that had pioneered the movement. This worrisome conflict, detailed fully in Grant's account of our work, initially focused on the ever delicate issue of whether field-workers disclose their notes to those on the site, but eventually implicated the FIPSE staff and their unhappiness with our approch to hard-core policy (Grant 1979a). Because the Alverno president was a figure of note in the highly interconnected world of higher education in the United States, I worried that news of these problems would filter out to Oregon. Such might undercut the precarious groundwork I had established and create time consuming demands from me for my notes and findings or explantions about the troubles at Alverno so that my theoretical footing and fieldwork approaches would be seriously corroded. I now realize I might have turned such a situation to my advantage and used it to clarify various individuals' definitions of competence-based education at Mt. Hood, but at the time, keenly aware of the episodic nature of the study, I could only see the threat to the delicate temporal pace and balancing of the data gathering and analysis.

If news of the problems with our study at Alverno reached Mt. Hood, it was never apparent to me and did not influence my fieldwork. That was, indeed, fortunate for me, for events erupted during a period in the summer when I was not at Mt. Hood that made my work unexpectedly difficult without the "fall out" I had anticipated from the crisis at Alverno. Those events wiped away most of the contacts I had worked so carefully on during the first year and necessitated that I begin anew to negotiate entree and to define the theoretical focus—again in an episodic mode.

What happened was that the long festering student resentment of the assessment mode of the competence-based nursing program led to a student rebellion. In the ensuing confrontation between the faculty and the administration the faculty resigned. By the time I learned of this from one of the faculty who had resigned and from one of my research team colleagues working at a nearby site in the state of Washington, all but one of the initial faculty had left and an entirely new set of instructors had been hired. The conflict at one point had even involved the college board of directors and the State Board of Nursing.

Though my notes of the spring had contained students' gripes about the modes and faculty comments about their awareness of the gripes, nothing in my notes would have led me to anticipate theoretically the rebellion nor the subsequent events. Nor did I find any hints when I reanalyzed those notes later. Perhaps my fieldwork relationships were more shallowly realized than I had thought, given the episodic nature of the work, and the students did not trust me enough to confide where they thought their discontent might take them. For their part, faculty were perhaps afraid to share with me their deep concerns about this part of an innovative program that was clearly very important to them, but there was sufficient trust in the relationship with the faculty for one of them to write me about the events.

In retrospect, I think theory and method and memories of past work with students, converged to blinker the work at that point. Theoretically, as I have already noted, I was primarily interested in the shaping of definitions as the innovation was implemented primarily, I thought, by the faculty. Intently focused on definitions, and recalling from the earlier study of student nurses that griping was a well-integrated part of their being-in-the world, I glossed the glimmerings of conflict and the fact that the students, as well as the faculty, were the implementors of the innovation. Because my fieldwork methodology was episodic, not continual and immersed as in the earlier study, I was unable to pick up the intervening steps and structural aspects between student griping and full blown rebellion, supported by the anti-faculty administration.

The fieldwork then took on much more complex dimensions: I had to warm up rather slender contacts in the administration to renegotiate entree, since the entire faculty was new. I had to establish myself with the new faculty, who happily for me, were welcoming and cordial. I had to dig out the story of the rebellious students, by then in their second year. And, though it did not please the college administration, I contacted and interviewed extensively the faculty members who had resigned and some members of the Oregon Board of Nursing. What had always been very demanding fieldwork in a limited time frame now became positively overwhelming. Fortunately for me, the human tendency to want to tell one's side of a study such as this eased many an interview and an interaction and the data, if anything, flowed much more easily into my notebook than they had in the difficult beginning days.

Disruptive events, be they interpersonal or institutional, yield harvests of data and insight that otherwise might be missed. What became apparent as I picked my way delicately back through the shards of early notes and the new data coming into my notebook was the part multiple actors played in implementing the innovative program: the dispossessed faculty, the new faculty, the rebellious students, the administration, the community college district and the State Board of Nursing. I could see that these various players each had some knowledge of the program and its implementation, but I also came to see that the conceptualization of the "social distribution of knowledge" (Schutz 1962, p. 350) had to be expanded to incorporate the relative power which holders of the definitions exercised. The powerful players here, apparently more than they themselves realized, were the students who gained administrative support, and the administration itself. Mindful of the newspaper stories in the conservative college district about the fuss with the students, stories that were apt to upset voters who would vote in future funding elections, the skittish administration refused requests that faculty had made after the rebellion and set conditions that the faculty could not accept and hence, resigned. Now, as then, my sympathies lay with faculty, perhaps influenced by my own sister's experiences in her own community college career and by my interpretations of the faculty's struggles and what I perceived to be a doltish college administration (Olesen 1979b).

Theoretically, then the question of power had to be linked to the issue of multiple definitions and players, some of whom, namely, the students, were clearly as much the implementors of the innovation as were the faculty. I also had to reconceptualize the innovation, competence-based education, not as a homogenous set of ideas, but as a parcel of issues that were differently defined, and one of which, the detested assignment system, would become the focus for rebellion and change.

Policy in the Mt. Hood Study

Theory and policy were closely intertwined from the very beginning in this study. I believed that I could best understand the history and phases of the nursing program at Mt. Hood by regarding it as an innovation around which different definitions held by various interacting parties would emerge. The theoretical framework of interactionism was ideally suited to this, but, as I have noted, proved blinkering at one point. Through it one could grasp the transformation of ideas that constituted the innovative nursing program. Consequently, when I presented my report on different occasions to the new faculty, the administration and members of the original faculty who had resigned, I did not deal with "outcomes," but, rather the processes of implementation around the innovation, the processes themselves being policy: "Social interactions in an institutional context produce educational innovation, but also generate a career for the innovation. If forceful and striking, that career may well transform the situation of individuals who actively contributed to the innovation, those who were implicated, and the institution of which all are apart. Moreover, such careers not only transcend the innovator and the institution, but live into new eras as a source for further change. At Mt. Hood, the innovators' dream became a drama with unwitting players, unforeseen scripts and occasionally unmanageable properties, all of which share with the innovators the credit or blame for the fate of the innovation" (Olesen 1979b, p. 362).

CONCLUSION

Invoking one's past as I have here is not the easiest of tasks: one anguishes over sins of omission and commission and recoils in embarrassment or guilt over images of the self in past fieldwork. Yet some unrepentant (or perhaps reborn) part of the field-worker self surges to the fore in consciousness and anticipates the next ethnographic adventure. What would I take to that adventure from these different studies as regards theory and policy, the themes of this volume?

Most definitely I believe that ethnography can be done in a variety of circumstances not usually embraced in the classical imagery inherited from anthropology and naturalistic sociology at the University of Chicago. As always, much, indeed everything, depends on what one wishes to know or to find out or to relate to theory in one's own field, but contemporary circumstances frame the possibilities for being, knowing, finding and relating in ways that do not permit immersion in the ways in which we have taken to be the sine qua non of our method. I would even here advance the heresy that the ideal conditions of immersion over time were not always fully met in every classical ethnography either in anthropology or in sociology and that our imagery of fieldwork *is* obdurately romanticized in spite of the steady progress to explicating the method in recent years. I for one celebrate that romanticized aspect of our work, for to me it speaks to the recognition in our work and our analyses of the untidiness of human endeavors, our own as field-workers, included.

Theory is a way of recognizing and understanding untidiness, or resolving it as the case may be. In my own work the symbolic interactionist perspective has never been absent from my fieldwork, but I cannot claim to have advanced that perspective. I think much that I have done has expanded aspects of that perspective, and when I venture into the field again, that perspective, plus the insights of cultural anthropology and historical sociology, would surely accommodate me.

As to policy, in spite of my recitation here that in at least two studies, though policy may have guided the study, the impact on policy was diffuse, I remain persuaded that ethnography has an important part to play in studies that relate directly to policy or on issues of interest to policymakers. There is a cognitive gap between depictions of populations statistically and the ways in which lives of members of that population are lived, even as there is a cognitive and interpretive gap between an institutionally approved policy and the ways in which those who are influenced by the policy will live their lives. Fieldwork can flesh out our picture of those lives. More cogently, the construction of policy as a process, amenable to theoretical and methodological strategies that focus on "the transformation of intentions" that is policy (Estes and Edmonds 1981) summons fieldwork to the fore as a useful method.

What perspective of policy and fieldwork as reconceptualized through the lens of "the transformation of intentions" provides is that clarification of issues, recognition of definitions *are* policy issues, and fieldwork is eminently suitable to analyze those. The three studies reviewed here, the nature of baccalaureate education, the experience of temporary service workers, and the implementation of an innovation all clarify policy issues in this sense. So, I would take to the next fieldwork venture my understandings that policy is a process, that its features can be redefined, and that one should not expect that one's research report will become a policy document, but, rather, that parts of that can clarify and move policy forward.

As I move forward to the next fieldwork venture with whatever complexities it poses for theoretical or policy relevant conduct and analysis, I would hope that my good fortune with respect to my co-researchers in the past would hold in that venture, as it did in the three studies reported here. These colleagueal ties, both with co-workers in the field and those others not present, are critical to my work for what they give in criticisms, support and inspiration, hence are particularly important in continual emergence of self as reflexive field-worker and, as noted at the outset of this paper the ongoing sedimentation of self.

Finally, reflecting on what I have attempted to recount and reconstruct in this chapter, I would, I think, consciously, attempt to attend to the properties of context and temporality in fieldwork, those themes that reflect structure and process that are integral to the art and craft of fieldwork, but which, for the most part, lie unexplicated.

REFERENCES

Ball, D.W.
1968 "Toward a Sociology of Telephones and Telephoners." Pp. 59-75 in M. Truzzi (ed.), *Sociology and Everyday Life*. Englewood Cliffs, NJ: Prentice-Hall.

Becker, H., Geer, B., Hughes, E.C., and Strauss, A.L.
1961 *Boys in White, Student Culture in Medical School.* Chicago: University of Chicago Press.
Blauner, R.
1964 *Alienation and Freedom.* Chicago: University of Chicago Press.
Bulmer, M.
1986 "The Value of Qualitative Methods." Pp. 180-204 in M. Bulmer (ed.), *Social Science and Public Policy.* London: Allen and Unwin.
Burgess, R.
1984 *In the Field.* London: Allen and Unwin.
Charmaz, K.C.
1980 "The Social Construction of Pity." Pp. 123-145 in N. Denzin (ed.), *Studies in Symbolic Interaction,* Vol. 3. Greenwich, CT: JAI Press.
Cook, T.D. and Reichardt, C.S.
1979 *Qualitative and Quantitative Methods in Evaluation Research.* Beverly Hills, CA: Sage.
Davis, F.
1968 "Professional Socialization as Subject Experience: The Process of Doctrinal Conversion among Student Nurses." In H.S. Becker et al. (eds.), *Institutions and the Person: Papers Presented to Everett Hughes.* Chicago: Aldine.
1972 "Rituals of Annunciation: On the Good Fortune of Getting Married, or Symbolism vs. Functionalism." Pp. 13-20 in F. Davis (ed.), *Illness, Interaction and the Self.* Belmont, CA: Wadsworth.
1979 *Yearning for Yesterday.* New York: Free Press
Davis, F. and Olesen, V.L.
1963 "Initiation into a Woman's Profession: Identity Problems in the Status Transition of Co-ed to Student Nurse." *Sociometry* 26 (March): 89-101.
Davis, F., Olesen, V.L., and Whittaker, E.W.
1966 "Problems and Issues in Collegiate Nursing Education." Pp. 138-175 in F. Davis (ed.), *The Nursing Profession.* New York: John Wiley.
Denzin, N.K.
1984 *On Understanding Emotion.* San Francisco: Jossey-Bass.
Emerson, R.
1983 *Contemporary Field Research.* Boston: Little Brown.
Estes, C.L. and Edmonds, B.C.
1981 "Symbolic Interaction and Social Policy Analysis." *Symbolic Interaction* 4(Spring):75-86.
Finch, J.
1986 *Research and Policy: The Uses of Qualitative Methods in Social and Educational Research.* London: Falmer.
Foster, G.M., Colson, E., Scudder, T., and Kemper, R.V.
1976 "Long-term Field Research in Social Anthropology." *Current Anthropology* 17(September):494-496.
Glaser, B.G.
1978 *Theoretical Sensitivity.* Mill Valley, CA: The Sociology Press.
Glaser, B.G. and Strauss, A.L.
1967 *The Discovery of Grounded Theory: Strategies for Qualitative Research.* Chicago: Aldine.
Goffman, E.
1956 "Embarrassment and Social Organization." *American Journal of Sociology* 62(November):264-271.
1959 *The Presentation of Self in Everyday Life.* Garden City, NY: Doubleday.

Grant, G.
1979a "Epilogue." Pp. 439-490 in G. Grant (ed.), *On Competence*. San Francisco: Jossey-Bass.
1979b "Prologue: Implications of Competence-Based Education." Pp. 1-17 in G. Grant (ed.), *On Competence*. San Francisco: Jossey-Bass.
Gross, E. and Stone, G.P.
1964 "Embarrasssment and the Analysis of Role Requirements." *American Journal of Sociology* 70(Spring):1-15.
Halfpenny, P.
1969 "The Analysis of Qualitative Data." *Sociological Review* 27(4):799-825.
Hammersley, M. and Atkinson, P.
1983 *Ethnography: Principles in Practice*. London: Tavistock.
Hochschild, A.R.
1984 *The Managed Heart*. Berkeley: University of California Press.
Hughes, E.C.
1984a "Cycles, Turning Points and Careers." Pp. 124-131 in E.C. Hughest (ed.), *The Sociological Eye*. New Brunswick, NJ: Transaction.
1984b "The Dilemmas and Contradictions of Status." Pp. 132-135 in E.C. Hughes (ed.), *The Sociological Eye*. New Brunswick, NJ: Transaction.
Lewin, E. and Damrell, J.
1978 "Female Identity and Career Pathways: Post-baccalaureate Nurses Ten Years After." *Sociology of Work and Occupations* 5:31-54.
Lewin, E. and Olesen, V.L.
1980 "Lateralness in Women's Work: New Views on Success." *Sex Roles: A Journal of Research* 6(Fall):619-629.
1985 "Occupational Health and Women: The Case of Clerical Work." Pp. 53-85 in E. Lewin and V.L. Olesen (eds.), *Women, Health and Healing: Toward a New Perspective*. London: Tavistock-Methuen.
Liebow, E.
1967 *Tally's Corner*. Boston: Little Brown.
Lofland, J. and Lofland, L.H.
1984 *Analyzing Social Settings*. Belmont, CA: Wadsworth.
Lofland, L.H.
1982 "Loss and Human Connection: An Exploration in the Nature of Social Bonds." Pp. 219-242 in W. Ickes and E. Knowles (eds.), *Personality, Roles and Social Behavior*. New York: Springer-Verlag.
Lynd, H.M.
1958 *On Shame and the Search for Identity*. New York: Harcourt Brace Jovanovich.
Manning, P.
1987 *Semiotics and Fieldwork*. Beverly Hills, CA: Sage.
McCall, G.J. and Simmons, J.L.
1966 *Identities and Interactions*. New York: Free Press.
McNally, F.
1979 *Women for Hire: A Study of the Female Office Worker*. London: Macmillan.
Merton, R.K., Reader, G.G., and Kendall, P.L. (eds.)
1957 *The Student Physician*. Cambridge: Harvard University Press.
Nelson, C.
1988 "An Anthropologist's Dilemma: Fieldwork and Interpretive Inquiry." *Aliph, Journal of Comparative Poetics* (Spring).

Olesen, V.L.
1968 "On the Possibility of Doing Research with Temporary Clerical Workers." Unpublished memo, Department of Social and Behavioral Sciences, University of California, San Francisco.
1973 "What Happens after Schooling: Notes on Post-institutional Socialization in the Health Occupations." *Social Science and Medicine* 7(1):61-75.
1974a "Temporary Clerical Workers and Occupation Integration: a Little Explored Sector of the Urban Struggle." Presentation to the American Psychiatric Association.
1974b "Transitory Urban Workers and Occupational Identity." Final Report Prepared for the Center for the Study of Metropolitan Problems, National Institutes of Health, U.S. Department of Health, Education and Welfare.
1977 "Musings, Some Serious, Some not on the Years of the FIPSE Field Work." Unpublished manuscript, Department of Social and Behavioral Sciences, University of California, San Francisco.
1979a "Federal Regulations, Institutional Review Boards and Qualitative Social Science Research." Pp. 45-55 in M.L. Wax and J. Cassell (eds.), *Federal Regulations, Ethical Issues and Social Research.* Boulder, CO: Westview.
1979b "Overcoming Crises in a New Nursing Program: Mt. Hood Community College." Pp. 335-362 in G. Grant (ed.), *On Competence.* San Francisco: Jossey-Bass.
1982 "Rethinking the 'Field': Conceptualization from a Study of Temporary Clerical Workers." Unpublished manuscript, Department of Social and Behavioral Sciences, University of California, San Francisco.
1983 "Episodic Fieldwork: Strategies and Questions." Unpublished manuscript, Department of Social and Behavioral Sciences, University of California, San Francisco.
1987 "Dynamics of Change in One's Profession and Self-Assessment: A Research Note on Aging Among Mid-life Women." Unpublished manuscript, Department of Social and Behavioral Sciences, University of California, San Francisco.

Olesen, V.L. and Katsuranis, F.
1978 "Urban Nomads: Women in Temporary Clerical Services." Pp. 63-72 in A.H. Stromberg and S. Harkess (eds.), *Women Working.* Palo Alto, CA: Mayfield.

Olesen, V.L. and Whittaker, E.W.
1966 "Adjudication of Student Awareness in Professional Socialization: The Language of Laughter and Silence." *The Sociological Quarterly* 7(Summer):381-96.
1968 *The Silent Dialogue: A Study in the Social Psychology of Professional Socialization.* San Francisco: Jossey-Bass.

Parlett, M. and Dearden, G.
1977 *Introduction to Illuminative Evaluation.* Cardiff-by-the-Sea, CA: Pacific Soundings.

Patton, M.Q.
1980 *Qualitative Evaluation Methods.* Beverly Hills, CA: Sage.

Phillipson, M.
1972 "Theory, Methodology and Conceptualization." Pp. 77-116 in P. Filmer, M. Phillipson, D. Silverman, and D. Walsh (eds.), *New Directions in Sociological Theory.* London: Collier-Macmillan.

Riesman, D. and Grant, G.
1978 *The Perpetual Dream: Reform and Experiment in the American College.* Chicago: University of Chicago.

Riesman, D., Gusfield, J., and Gamzon, Z.
1970 *Academic Values and Mass Education.* Garden City, NY: Doubleday.

Scheff, T.
1983 "Toward Integration in the Social Psychology of Emotion." *Annual Review of Sociology* 9:333-354.

Scheler, M.
1961 "On the Phenomenology and Sociology of Ressentiment." Pp. 43-79 in L. Coser (ed.), *Ressentiment.* Glencoe, IL: The Free Press.

Scholte, B.
1961 "Toward a Reflexive and Critical Anthropology." Pp. 430-443 in D. Hymes (ed.), *Reinventing Anthropology.* New York: Vintage.

Schutz, A.
1962 *Collected Papers, Vol. I, The Problem of Social Reality,* M. Nathanson (ed.). The Hague: Martinus Nijhoff.

Silverman, D.
1985 *Qualitative Methodology and Sociology.* London: Gower.

Simmel, G.
1950 "How is Society Possible?" Pp. 337-356 in K. Wolff (ed.), *Georg Simmel.* Columbus: Ohio State University.

Spradley, J.P.
1979 *The Ethnographic Interview.* New York: Holt, Rinehart, Winston.

Stanley, L. and Wise, S.
1983 *Breaking Out: Feminist Consciousness and Feminist Research.* London: Routledge and Kegan Paul.

Strauss, A.L. (ed.)
1946 *The Social Psychology of George Herbert Mead.* Chicago: University of Chicago Press.
1959 *Mirrors and Masks: The Search for Identity.* Glencoe, IL: The Free Press.

Whittaker, E.W.
1986 *The Mainland Haole, the White Experience in Hawaii.* New York: Columbia University Press.

NOTES ON CONTRIBUTORS

Jacquelin Burgess has been a lecturer in the Department of Geography, University College, London, England, since 1975. She was appointed to teach and research in the field of environmental perception but over the last twelve years, her work has evolved from a predominantly psychological orientation to a more contextual appreciation of the social and cultural contexts within which meanings and values for landscapes and places are shaped. Currently she teaches courses in cultural geography and environmental interpretation. She is joint editor (with John Gold) of: *Valued Environments* (Allen and Unwin, 1982) and *Geography, the Media and Popular Culture* (Croom Helm, 1985).

Robert Burgess is Professor of Sociology and Director of CEDAR (Centre for Educational Development, Appraisal and Research) at the University of Warwick, England. His main teaching and research interests are in social research methodology (especially qualitative methods) and the sociology of education (especially the study of schools, classrooms and curricula). He is currently writing an ethnographic restudy of a comprehensive school on which he has already published several papers. His main publications include: *Experiencing Comprehensive Education* (1983), *In the Field: An Introduction to Field Research* (1984), *Education, Schools and Schooling* (1985), and *Sociology, Education and Schools* (1986) together with fourteen edited volumes on qualitative methods and education. He is President of the British Sociological Association.

Joan Chandler is currently Lecturer in Sociology and Nurse Tutor for Plymouth Health Authority. She was previously a post-doctoral Research Fellow in the Department of Social and Political Studies at Plymouth Polytechnic, England working on a project investigating aspects of the Youth Training Scheme. She recently completed a Ph.D. that was concerned with a study of wives of members of the Royal Navy in the United Kingdom.

Janet Finch is Professor of Social Relations at the University of Lancaster, England where she teaches social policy. She has a background in sociology and is a former Chair of the British Sociological Association. In 1988 she became joint editor of the journal *Sociology.* Her main research and publications are in the fields of family, gender and social policy; also in education. She has published on research methods, including the book *Research and Policy: The Uses of Qualitative Methods in Social and Educational Research* (Falmer Press, 1986).

Stephen Fox is a Lecturer in the Centre for the Study of Management Learning at the University of Lancaster, England. He has recently completed a Ph.D. on students who were studying on the part-time Master of Business Administration (MBA) Degree at Manchester Business School in England.

Norman L. Friedman is Professor of Sociology at California State University, Los Angeles. He previously taught at the University of Southern California and the University of Missouri, and has held research fellowships at the Kansas City Institute for Community Studies and at Brandeis University. He has published over forty articles in various journals, magazines and research annuals, mainly about occupations, ethnicity, mass media and education.

Barrie Goldsmith joined University College London, England to teach the M.Sc. Course in Conservation in 1967 and has recently become a Senior Lecturer. His principal interests are plant ecology, conservation evaluation, and recreation management. He has worked as a forest ecologist in Nova Scotia, Canada, for three years in 1976-1979 and again in 1986 when he attempted to select a series of forest reserves. He has co-edited two books on conservation, *Conservation in Practice* (1974) and *Conservation in Perspective* (1983) and is on the Editorial Boards of *Biological Conservation* and *Landscape & Urban Ecology.*

Carolyn Harrison lectures in biogeography and resource management in the Department of Geography at University College, London, England. Her early work focused on the ecological impact of recreation on various semi-natural habitats and more recently has examined the critical role managers play in reconciling the recreational use of open spaces with nature conservation objectives. She has taught on the postgraduate MSc. in Conservation for over fifteen years and has published various articles in scientific and geographical journals.

James M. Henslin received a B.D. from Concordia Seminary in St. Louis, Missouri, and both his Masters and Ph.D. degrees from Washington University, also in St. Louis. He began his teaching career at the University of Missouri at St. Louis and is currently Professor of Sociology at the Edwardsville campus of Southern Illinois University. His primary interests are social problems, deviance, and marriage and the family. He has contributed to the *American Journal of Sociology* and *Social Problems,* and authored or edited *Down to Earth Sociology* (fifth edition), *Marriage and Family in a Changing Society* (third edition in preparation), *Social Problems* (with Donald Light), *The Sociology of Sex* (with Ed Sagarin), and *Introducing Sociology.* His current research focus is the homeless, for which he has travelled to skid rows across America and about which he now is writing a book, as well as editing another.

Jennifer Mason is currently Lecturer in Social Policy in the Department of Social Administration, University of Lancaster, England, where she was previously Research Officer on the Family Obligations Project. Her research interests include the sociology of gender, family and kinship, and the use of qualitative methodology. She has conducted research on gender inequalities in long-term marriages, and has published several articles on the topic.

Kristine Mason is a senior lecturer in the sociology of education at the College of St. Paul and St. Mary, Cheltenham, England. Previously she has lectured at Isleworth and Brighton Polytechnics and also in Buea, Cameroon. Her main professional and research commitments and interests are directed toward equal opportunities and education. She completed her Ph.D., "Schoolgirls in a Rural Context," at the University of Aston in Birmingham, England, in 1987, and has published some of the material from this research in "Schooling in Rural England—Jobs for the Girls?" in *Schooling in Turmoil,* edited by G. Walford (1985). More recently she has been involved in a collaborative ethnographic research project in the School of Secondary Education at the College of St. Paul and St. Mary which has focused upon a Certificate of Prevocational Education program in two Gloucestershire schools in England.

Virginia Olesen is professor of sociology and co-director of the Women, Health and Healing Program in the Department of Social and Behavioral Sciences, School of Nursing, University of California, San Francisco. She is concluding work on the phenomenology of self-care and is starting research with the mid-life, mid-career women studied twenty five years ago in fieldwork described in her essay. She is also collaborating in studies of socialization of ethical behavior in student nurses and of emergent professional identity in university nursing students in the People's Republic of China.

Alan Prout is currently Lecturer in Sociology at the University of Keele, England. He was previously Senior Lecturer in Health Studies at the South Bank Polytechnic, London, England. Before this he was a teacher in further education and then Research Associate, first at the Health Education Studies Unit and later at the Child Care and Development Group, both at the University of Cambridge. His doctoral work was done at the Centre for Medical Social Anthropology at Keele University. His research has brought together the sociologies of health and schooling and he has published work in both fields. He is also active in the development of a sociology and ethnography of childhood.

Helen Rainbird trained as a social anthropologist in the University of Durham and conducted fieldwork in Peru. Subsequently she has conducted research in a variety of settings in the United Kingdom. At the present time she is a Senior Research Fellow in the Industrial Relations Research Unit at the University of Warwick, where she was previously a Research Fellow in the Institute for Employment Research.

AUTHOR INDEX

SUBJECT INDEX

Studies in Qualitative Methodology

Edited by **Robert G. Burgess,** *Department of Sociology, University of Warwick*

The last decade has witnessed a considerable increase in research that could be broadly described as ethnographic, qualitative or a case study among investigators working within such disciplines and areas of study as sociology, criminology and education, as well as sub-fields like industrial relations and the sociology of health and healing. Such work draws on a style of investigation traditionally used by social anthropologists and includes methods such as participant observation, unstructured interviews and documentary evidence. This range of research methods is commonly included under the term field research and qualitative methodology.

It is the intention of these research annuals on qualitative research to take up issues and debates in this area that relate to: methodology, the relationship between data collection and data analysis, the relationship between theory and mehtod and the implications of qualitative research for social policy and evaluation.

Volume 1, Conducting Qualitative Research
1988, 257 pp. $58.50
ISBN 0-89232-762-6

This volume contains a range of papers that are broadly concerned with the conduct of qualitative research. There are three papers concerned with case studies, by Jennifer Platt, Louis Smith and Tom Schuller. There are two papers that contain reflections on research experience by Rebecca Klatch and Joan Cassell. A further three papers deal with interviewing and with the analysis of transcriptions. These papers are by George Moyser, Robert Burgess and Barbara Rawlings. There is a paper on the implications of data protection legislation for qualitative research by Anne Ackroyud and a paper on reflexivity by Peter Kloos.

JAI PRESS

JAI PRESS

ity, *Rebecca E. Klatch, University of California, Santa Cruz.* **The Relationship of Observer to Observed When Studying Up,** *Joan Cassell, Washington University.* **Non-Standarized Interviewing in Elite Research,** *George Moyser, University of Vermont.* **Conversations with a Purpose: The Ethnographic Interview in Educational Research,** *Robert G. Burgess, University of Warwick, England.* **Local Knowledge: The Analysis of Transcribed Audio Materials for Organizational Ethnography,** *Barbara Rawlings, University of Manchester, England.* **Ethnography, Personal Data and Computers: The Implications of Data Protection Legislation for Qualitative Social Research,** *Anne V. Akeroyd, University of York, England.* **No Knowledge Without A Knowing Subject,** *Peter Kloos, University of Leiden.* **Notes on Contributors. Name Index. Subject Index.**

Volume 2, Reflections on Field Experience
1990, 244 pp. $58.50
ISBN 1-55938-023-3

CONTENTS: Preface. Introduction, *Robert G. Burgess.* **Becoming an Ethnomethodology User: Learning a Perspective in the Field,** *Stephen Fox, University of Lancaster.* **Decision Taking in the Fieldwork Process: Theoretical Sampling and Collaborative Working,** *Janet Finch and Jennifer Mason, University of Lancaster.* **"It's Not a Loverly Place to Visit, and I Wouldn't Want to Live There",** *James A. Henslin, Southern Illinois University.* **Expectations and Revelations: Examining Conflict in the Andes,** *Helen Rainbird, University of Warwick.* **Not Waving, But Bidding: Reflections on Research in a Rural Setting,** *Kristine Mason, College of St. Paul and St. Mary.* **Research and the Relevance of Gender,** *Joan Chandler, Plymouth Polytechnic.* **Pale Shadows for Policy: Reflections on the Greenwich Open Space Project,** *Jacquelin Burgess and Carolyn M. Harrison, University College, London.* **Comment: Can Popular Values Be Derived from 33 People?,** *F.B. Goldsmith, University College, London.* **"Spitting Image: 'Pure' and 'Applied' Research in the Culture of Sociology",** *Allan Prout, South Bank Polytechnic.* **Conventional Covert Ethnographic Research By a Worker: Considerations from Studies Conducted as a Substitute Teacher, Hollywood Actor, and Religious School Supervisor,** *Norman L. Friedman, California State University, Los Angeles.* **Immersed, Amorphous and Episodic Fieldwork: Theory and Policy in Three Contrasting Contexts,** *Virginia Olesen, University of California, San Francisco. Notes on Contributors.*

Volume 3, In preparation, Summer 1991
ISBN 1-55938-246-5 Approx. $58.50

JAI